Geopolymers

Geopolymers presents a complex and interdisciplinary study in the fields of physics, chemistry, materials science, and civil engineering on oxide materials based on mineral wastes, known as geopolymers.

Considering the ideal requirements for developing eco-friendly materials for industrial applications, this book describes how to design and develop different types of geopolymers that use mineral wastes or natural aluminosilicates as raw materials. It contains advanced knowledge and information regarding geopolymer manufacturing, development, characterization, and applications in soil stabilization, civil engineering, or ceramics.

This book is relevant for undergraduate and graduate students conducting fundamental and applied research in the fields of materials engineering, ceramics engineering, and water processing.

Geopolymers
Properties and Applications

Edited by
Petrica Vizureanu,
Mohd Mustafa Al Bakri Abdullah,
Rafiza Abdul Razak, Dumitru Doru Burduhos-Nergis,
Liew Yun-Ming, and Andrei Victor Sandu

CRC Press
Taylor & Francis Group
Boca Raton London New York

CRC Press is an imprint of the
Taylor & Francis Group, an **informa** business

Designed cover image: Optical micrography 50X of fly ash-red mud blended geopolymer, obtained by the editors.

First edition published 2024
by CRC Press
6000 Broken Sound Parkway NW, Suite 300, Boca Raton, FL 33487-2742

and by CRC Press
4 Park Square, Milton Park, Abingdon, Oxon, OX14 4RN

CRC Press is an imprint of Taylor & Francis Group, LLC

Library of Congress Cataloging-in-Publication Data
Names: Vizureanu, Petrica, editor.
Title: Geopolymers : properties and applications / edited by Petrica Vizureanu, Mohd Mustafa Al Bakri Abdullah, Rafiza Abdul Razak, Dumitru Doru Burduhos-Nergis, Liew Yun-Ming, and Andrei Victor Sandu.
Other titles: Geopolymers (CRC Press)
Description: First edition. | Boca Raton : CRC Press, [2024] |
Includes bibliographical references and index.
Identifiers: LCCN 2023019710 (print) | LCCN 2023019711 (ebook) |
ISBN 9781032486710 (hbk) | ISBN 9781032486727 (pbk) | ISBN 9781003390190 (ebk)
Subjects: LCSH: Inorganic polymers.
Classification: LCC TA455.P583 G46 2024 (print) | LCC TA455.P583 (ebook) |
DDC 620.1/92–dc23/eng/20230427
LC record available at https://lccn.loc.gov/2023019710
LC ebook record available at https://lccn.loc.gov/2023019711

ISBN: 978-1-032-48671-0 (hbk)
ISBN: 978-1-032-48672-7 (pbk)
ISBN: 978-1-003-39019-0 (ebk)

DOI: 10.1201/9781003390190

Typeset in Times
by codeMantra

Contents

Chapter 3   Geopolymer-Reinforced Steel Fibers... 42

Meor Ahmad Faris Meor Ahmad Tajudin,
Nurul Aida Mohd Mortar, Muhammad Faheem Mohd Tahir,
Dumitru Doru Burduhos-Nergis, and Poppy Puspitasari

Chapter 4   Geopolymer Lightweight Aggregate... 52

Rafiza Abdul Razak, Alida Abdullah,
Mohd Mustafa Al Bakri Abdullah, Petrica Vizureanu, Eva Arifi,
and Andrei Victor Sandu

Preface

Currently, there is a continuous concern for research and development of eco-friendly materials. Thus, the researcher focuses on developing new technologies for the synthesis of advanced materials that are based on recycled raw materials and exhibit comparable or even higher properties than conventional ones. Therefore, a new class of oxide materials, known as geopolymers, that exhibit excellent durability and significantly lower carbon footprint than those based on ordinary Portland cement have emerged.

Geopolymers are inorganic materials that have been introduced in the literature, especially due to Joseph Davidovits' works back in 1970. However, their applications seem to date back to the Romans in the construction of Colosseum and Pantheon, or even earlier, back to 2700 B.C. in Egyptian Pyramids construction. These materials were used and developed without knowing the term geopolymers, a term introduced about 50 years ago that boosted the development of these products.

Considering that geopolymers are obtained by mixing a solid material rich in aluminum and silicon oxides with a strong alkaline solution, their properties strongly depend on the characteristics of the raw materials. To ideally meet the requirements for developing eco-friendly materials with potential for industrial applications, in general, the target material must exhibit simultaneously: (i) comparable (similar or higher) properties to those of conventional materials; (ii) raw materials available in large quantities and distributed worldwide; (iii) possibility of using waste as raw materials; (iv) simple/cheap manufacturing technology (equipment and personnel training); and (v) product versatility (the geopolymer will possess suitable characteristics for multiple applications, not just the construction sector). The first requirement is often met in the case of geopolymers. However, the last four requirements must be customized.

Population growth, urbanization, and large-scale consumption of raw materials are causing vast environmental, social, and economic problems across the globe. Although climate change is perhaps the most pervasive of these issues to be found in scientific and political discourse, there are many other issues associated with waste disposal, contamination, and land use that are closely intertwined with the consequences of humanity's expansion into the natural world. Therefore, the strong need for construction materials leads to a constantly growing market in this field.

This book will have a strong impact on all scientists, convincing them that the most used material, after water, namely ordinary Portland cement (OPC)-based materials, has an eco-friendly option. Therefore, most of the negative environmental effects associated with OPC production and use can be eliminated.

This book has 11 chapters that contain a complex and interdisciplinary study in the fields of physics, chemistry, materials science, and civil engineering on oxide materials based on mineral wastes, known as geopolymers. Also, it contains advanced knowledge and information regarding geopolymers manufacturing, development, characterization, and applications in soil stabilization, civil engineering, or ceramics.

This book is relevant for fundamental and applied research in the field of materials engineering because it shows the technology of eco-friendly geopolymers with industrial applications. It also describes how to design and develop different types of geopolymers that use mineral wastes or natural aluminosilicates as raw materials. Therefore, this book is addressed to M.Sc. students, Ph.D. students, industrial users from the fields of materials science, ceramics engineering, and water processing/cleaning, as well as researchers worldwide.

About the Editors

Petrica Vizureanu
Professor Ph.D. Eng.
Head of department at Department of Technology and Equipment for Materials Processing
Faculty of Materials Science and Engineering, "Gheorghe Asachi" Technical University of Iasi
Member of Technical Sciences Academy of Romania

Petrica Vizureanu is a full professor and researcher at the "Gheorghe Asachi" Technical University of Iasi, having more than 30 years of experience. He defended his PhD in 1999 in the field of materials science and engineering, and since 2010, he has been a PhD supervisor in the same field. He has over 300 publications, of which 240 are indexed in the Clarivate Web of Science. He has extensive experience in the fields of composite materials, ceramic materials, insulating materials, and optimization of material characteristics. The H-index is 24.

Professor Dr. Mohd Mustafa Al Bakri Abdullah, PTech, MMSET, AAE
Faculty of Chemical Engineering & Technology
Universiti Malaysia Perlis (UniMAP)
Malaysia
Vice President, World Invention Intellectual Property Associations (WIIPA)
Honorary Secretary—Malaysian Research & Innovation Society (MyRIS)

Mohd Mustafa Al Bakri Abdullah is a Professor and Deputy Vice Chancellor (Student Affairs and Alumni) at Universiti Malaysia Perlis (UniMAP). He specializes in geopolymer concrete, green concrete, and various other types of concrete. He has held various positions at UniMAP, including Dean, Faculty of Chemical Engineering and Technology, Director, Research Management Centre, and some more. He received awards and recognition for his geopolymer research. He also established the Center of Excellence in Geopolymer and Green Technology at UniMAP, which is a global leader in the field of geopolymers. He has engaged in research collaborations with institutions both within and outside the country and has an impressive research record, including numerous journal publications (750), books (30), and patents (35). He has also been appointed as a Visiting Professor at several universities and holds Associate Researcher positions at several international universities.

Assoc. Prof. Dr. Rafiza Abdul Razak, Malaysia
B.Eng. (Civil Engineering), M.Sc. (Structural Engineering),
Ph.D. (Materials Engineering)
Centre of Excellence Geopolymer and Green Technology
(CEGeoGTech), UniMAP Malaysia

Rafiza Abdul Razak is a research fellow at the Centre of Excellence in Geopolymer and Green Technology (CEGeoGTech), University of Malaysia Perlis (UniMAP). She is an Associate Professor at the Faculty of Civil Engineering Technology at UniMAP. As a Fellow of the Centre of Excellence Geopolymer and Green Technology (CEGeoGTech), Universiti Malaysia Perlis (UniMAP), she has actively been promoting the projects involving producing artificial lightweight aggregate by inclusive geopolymer and utilizing recycled material to produce lightweight aggregate, funded by the Ministry of Higher Education Malaysia, Fundamental Research Grant Scheme (FRGS). She has given numerous invited/keynote talks at international conferences, including a webinar organized by RAEng Frontiers Champion in 2021 that focused on recycled aggregate. She has also published 80 papers, including an impact journal, conference and proceeding papers, book chapters, and as a main author and co-author with an h-index of 19, many of which are related to the field of geopolymer. She also has a granted patent in this field. She was awarded the title of young scientist by UniMAP for 2 years in a row (2017 and 2018) due to her high publications and research awards below the age of 40. She was winner of "Double Gold Award" and "Special Recognition Award 2019" in Global Women Innovators Network Awards (GWIIN) in London, UK.

Dumitru Doru Burduhos-Nergis
Head of works (Lecturer) at Faculty of Materials Science and Engineering, "Gheorghe Asachi" Technical University of Iasi, Romania
Personal webpage: http://www.afir.org.ro/ddbn/

Dumitru Doru Burduhos-Nergis, Ph.D., Eng., is a materials engineering researcher with many years of experience in the field of geopolymers and eco-friendly materials. He has many research articles published in high-ranked journals and many books and chapters published by international publishing houses. The H-index is 9. He is a reviewer or guest editor for highly ranked journals such as *Molecules* (ISSN: 1420-3049), *Sustainability* (ISSN: 2071-1050), *Journal of Composites Science* (ISSN: 2504-477X), *Archives of Metallurgy and Materials* (ISSN: 1733-3490), *Advances in Materials* (ISSN: 2327-252X), *Applied Sciences* (ISSN: 2076-3417), and *Coatings* (ISSN: 2079-6412). He is also a member of the organizing committee or scientific committee of international conferences and congresses in the fields of materials engineering, materials science, and chemistry.

Dr. Liew Yun-Ming is currently serving as an Associate Professor in the Faculty of Chemical Engineering and Technology at Universiti Malaysia Perlis (UniMAP) and is also a senior researcher at the Center of Excellence in Geopolymer and Green Technology (CEGeoGTech). Her main fields of research include geopolymers, cement and concrete, and recycled materials. She has published 30 impact journals, 3 book chapters, and 55 conference papers, with a Scopus h-index of 19 and a total citation count of 1762. She is a leader and co-leader for several research grants provided by the Ministry of Higher Education, Malaysia, with a value of RM 271,200 (USD 62,380). At the same time, she is also a co-leader in international grants (RM 2.5 million/USD 564,670) provided by the European Union (EU) under Horizon 2020. Besides, she has also been appointed as a reviewer for distinguished journal papers, a technical committee member of internal conferences, an editorial board member for several multidisciplinary international journals, and a juror for national and international exhibitions.

Andrei Victor Sandu

Associate Professor at Faculty of Materials Science and Engineering, "Gheorghe Asachi" Technical University of Iasi, Romania

Senior Researcher at National Institute for Research and Development for Environmental Protection INCDPM

President of Romanian Inventors Forum

http://afir.org.ro/sav/

Dr. Eng. Andrei Victor Sandu is an Associate Professor at the Faculty of Materials Science and Engineering at Technical University "Gheorghe Asachi" of Iași. He has a PhD in Materials Engineering since 2012 with summa cum laude. He has published over 380 scientific articles, of which over 350 are indexed by Scopus and more than 300 are indexed by ISI Web of Science. The H-index is 26. He is the co-author of 31 patents and 9 other patent applications (Romania, R. Moldova and Malaysia), and he has published 9 books, 2 of which in the USA. He is the publishing editor for *International Journal of Conservation Science* (Web of Science and Scopus indexed) and *European Journal of Materials Science and Engineering.* He is also a reviewer for more than 20 Web of Science-indexed journals. He is a Visiting Professor at Universiti Malaysia Perlis and also President of Romanian Inventors Forum. Based on his expertise, he is also a senior researcher for National Institute for Research and Development for Environmental Protection (INCDPM) and a representative for Romania at IFIA (International Federation of Inventors' Associations) and WIIPA (World Invention Intellectual Property Associations).

Contributors

Alida Abdullah
Universiti Malaysia Perlis
Perlis, Malaysia

Mohd Mustafa Al Bakri Abdullah
Universiti Malaysia Perlis
Perlis, Malaysia

Romisuhani Ahmad
Universiti Malaysia Perlis
Perlis, Malaysia

Eva Arifi
Universitas Brawijaya
Malang, Indonesia

Dumitru Doru Burduhos-Nergis
Gheorghe Asachi Technical University
Iasi, Romania

Laila Mardiah Deraman
Universiti Malaysia Perlis
Perlis, Malaysia

Ratna Ediati
Institut Teknologi Sepuluh Nopember
Surabaya, Indonesia

Ooi Wan-En
Universiti Malaysia Perlis
Perlis, Malaysia

Ng Hui-Teng
Universiti Malaysia Perlis
Perlis, Malaysia

Kamarudin Hussin
Universiti Malaysia Perlis
Perlis, Malaysia

Wan Mastura Wan Ibrahim
Universiti Malaysia Perlis
Perlis, Malaysia

Thanongsak Imjai
Walailak University
Nakhon Si Thammarat, Thailand

Liyana Jamaludin
Universiti Malaysia Perlis
Perlis, Malaysia

Nur Ain Jaya
Universiti Malaysia Perlis
Perlis, Malaysia

Wei-Hao Lee
National Taipei University of
    Technology
Taipei, Taiwan

Liew Yun-Ming
Universiti Malaysia Perlis
Perlis, Malaysia

Nurul Aida Mohd Mortar
Universiti Malaysia Perlis
Perlis, Malaysia

Lim Jia-Ni
Universiti Malaysia Perlis
Perlis, Malaysia

Poppy Puspitasari
Universitas Brawijaya
Malang, Indonesia

Azmi Rahmat
Universiti Malaysia Perlis
Perlis, Malaysia

Shamala A. P. Ramasamy
Universiti Malaysia Perlis
Perlis, Malaysia

Rafiza Abdul Razak
Universiti Malaysia Perlis
Perlis, Malaysia

Puput Risdanareni
Universitas Brawijaya
Malang, Indonesia

Andrei Victor Sandu
Gheorghe Asachi Technical University
Iasi, Romania

Liyana Ahmad Sofri
Universiti Malaysia Perlis
Perlis, Malaysia

Muhammad Faheem Mohd Tahir
Universiti Malaysia Perlis
Perlis, Malaysia

Meor Ahmad Faris Meor Ahmad Tajudin
Universiti Malaysia Perlis
Perlis, Malaysia

I Nyoman Arya Thanaya
Udayana University
Bali, Indonesia

Petrica Vizureanu
Gheorghe Asachi Technical University
Iasi, Romania

Ong Shee-Ween
Universiti Malaysia Perlis
Perlis, Malaysia

Tee Hoe-Woon
Universiti Malaysia Perlis
Perlis, Malaysia

Heah Cheng-Yong
Universiti Malaysia Perlis
Perlis, Malaysia

Hang Yong-Jie
Universiti Malaysia Perlis
Perlis, Malaysia

Ng Yong-Sing
Universiti Malaysia Perlis
Perlis, Malaysia

Farah Farhana Zainal
Universiti Malaysia Perlis
Perlis, Malaysia

Khairunisa Zulkifly
Universiti Malaysia Perlis
Perlis, Malaysia

1 Bibliometric Analysis of Research Trends in Geopolymers

Dumitru Doru Burduhos-Nergis,
Petrica Vizureanu, and Andrei Victor Sandu
Gheorghe Asachi Technical University

Rafiza Abdul Razak and Romisuhani Ahmad
Universiti Malaysia Perlis

1.1 INTRODUCTION

Population expansion, urbanization, and increased raw material use are causing major environmental, social, and economic issues throughout the world (Vizureanu and Burduhos Nergis 2020; Dong et al. 2019). Climate change is undoubtedly the most widespread of these topics in scientific and political debate. However, there are other challenges related to waste disposal, pollution, and land usage that are inextricably linked to the repercussions of human expansion into nature. In order to counteract these negative impacts, the international community established the objective of sustainable development (Albino et al. 2009; Geissdoerfer et al. 2017). Sustainable development is primarily about ensuring that we satisfy the demands of the present without jeopardizing our ecosystem's ability to meet the requirements of future generations.

After water, the most often utilized material is ordinary Portland cement (OPC) (Shamsaei et al. 2021). OPC is widely involved in the building sector, and its use has a significant environmental impact. The production of OPC not only requires a large amount of virgin raw materials and energy, but it also produces considerable greenhouse gas emissions (Hao et al. 2022). One ton of OPC uses around 1.5 tons of raw materials and emits around 1 ton of CO_2 into the atmosphere (Shah et al. 2020). The cement sector alone is expected to contribute 7%–9% of global CO_2 emissions generated (Horan et al. 2022; He et al. 2019). Another disadvantage of OPC is that it may lack important qualities for some applications, such as quick mechanical strength development and chemical resistance (Söylev and Özturan 2014; Ahmad et al. 2020).

Recently, geopolymers have been of great research interest as the ideal OPC alternative for sustainable development (Samuvel Raj et al. 2023; Khalifeh et al. 2018). As geopolymer materials are made from industrial by-products, they do not require any energy from fossil fuels in the production process, making them an environmentally friendly alternative to traditional cement and concrete. Furthermore, geopolymer materials have been found to have higher compressive strength and lower water absorption when compared to OPC, making them more durable and resistant to environmental conditions (Ul Rehman et al. 2020). Due to their green qualities and superior mechanical performance, the geopolymers are a promising alternative to ordinary Portland cement (OPC) in the construction industry, which is in critical need of environmentally and energy-efficient building materials (Li et al. 2022). In comparison to OPC, geopolymer substitution might reduce current human-caused carbon emissions by 5%–7% (Davidovits 2015). OPC manufacturing is now the poorest performing industry in terms of CO_2 emissions to the atmosphere. Additionally, the three-dimensional molecular network of the geopolymer increases acid resistance, high-temperature stability, and early strength development (Burduhos Nergis et al. 2018; Jat et al. 2023). These materials are suitable for future architectural applications due to their improved performance and green possibilities.

1.2   STATE-OF-THE-ART IN GEOPOLYMERS

As the geopolymer topic has grown in popularity, it is critical to conduct a bibliometric analysis of the published literature to keep readers and policymakers informed about current and future research hotspots. There have been a few reviews and bibliometric studies on geopolymer, but no bibliometric research has been done using geopolymer in a broader sense of search strings that provide a larger number of bibliographic data to offer an unbiased and less subjective image of trends and developments. Currently, there is no clear picture as to which aspects of geopolymer research are advanced and which ones are not. Therefore, this section will show the beginning, trends, and progress of the most notable geopolymer research publishing journals, keywords, authors, and papers, starting with the first publication in this field and up to the current date. The bibliometric analysis of the current study provides important statistical insights into geopolymer development and use in the industry, emphasizing current and future research trends. Furthermore, it provides a deeper knowledge of the theoretical structure and core topics of geopolymers, as well as emphasizing critical concerns and significant contributions made to the growth of this discipline by top publications and writers in the field.

The most accurate and effective method for examining and analyzing the enormous amount of scientific material in the area of geopolymers is bibliometric analysis using various software programs. In this subject, the network visualization of keywords and authors of published articles is especially important. As a result, in this chapter, a brief presentation of the research evolution of publications, indexed by the Scopus database, in the field of geopolymers was conducted using the Bibliometrix software. The Bibliometrix software, on the other hand, can do bibliometric analysis and create data matrices for co-citation, coupling, scientific cooperation analysis, and co-word analysis. Furthermore, at the intersections of structural and temporal evolution, such as network analysis, factorial analysis, and theme mapping, new information emerges.

This study used a vast data mining approach involving a scientometric review of the literature and an in-depth discussion of the results to describe the current state-of-the-art in the geopolymers field. As this research domain constantly expands, scholars may experience information overload, which may stymie constructive research efforts and academic collaboration. As a result, a plan that allows the researcher to gather important data from the most dependable sources must be established and implemented. Because of the inherent subjective biases in literature reviews, scientometric approaches can help mitigate this shortcoming through the use of a computational tool.

To retrieve the data from the database, one analysis was done for the term "geopolymer," which is the most general keyword used to describe the geopolymer field. The Scopus database states that at the time of inquiry, a search for "geopolymer" yielded 10.583 results from all publications indexed in Scopus. Furthermore, different groups of keywords were created by processing these findings using the Bibliometrix software, and their applicability was screened based on various factors as described below. The data exported from Scopus on January 20, 2022 were processed using R version 4.2.2 the "Innocent and Trusting" free software environment for statistical computing and graphics. A brief description of the software and its accuracy is presented in the study by Aria and Cuccurullo (2017).

The presented chapter approaches the review of relevant literature in the field of geopolymers utilizing the bibliometric evaluation technique, which allows for the creation of visual presentations by statistically assessing the published literature. This technique allows you to highlight the distinct correlations between the keywords particular to each field of study and confirm the trend in each sector of research. The study was based on papers that were indexed in the Scopus database, which is one of the most selective in terms of publication rating and indexing. The main information about the retrieved data is presented in Table 1.1. As can be seen, the first publications in this field were identified in a paper released online in 1984. Also, all indexed papers are published in 1,351 sources by close to 16,000 authors. Another important aspect that can be observed from this data is the high annual growth rate of almost 15%, which shows an increasing interest in this field. A brief analysis of the types of documents shows that almost 70% of the publications are research papers (articles), while the rest is mostly divided between conference papers, book chapters, and reviews.

1.2.1 ANNUAL EVOLUTION OF PUBLISHED PAPERS IN THE FIELD OF GEOPOLYMERS

Figure 1.1 depicts the annual research output, which shows the evolution of interest in geopolymers. The first publication in the field of geopolymers appeared in 1984 with the work of J. Davidovits; when searching for the term "geopolymer" in the Scopus database, we can see that the number of publications continued to be small until 2006; after this year, the number of publications presented an exponential increase, reaching over 850 in 2019 and over 1,900 in 2022. Even though the overall growth rate is calculated at 14.75, if we remove the first 20 years, i.e., 1984–2023, the growth rate decreases to 13.85, despite the fact that almost 99% of the total papers were published in the last 20 years, i.e., between 2004 and 2023.

TABLE 1.1

Main Information about the Current Literature in the Field of Geopolymers Indexed by Scopus

Classification by	Result	Description	Result
General		Document type	
Timespan	1984:2023	Article	7,324
Sources (journals, books, etc.)	1,351	Article in press	2
Documents	10,583	Book	9
Annual growth rate (%)	14.75	Book chapter	261
Document average age	4.89	Conference paper	2,277
Average citations per doc	23.52	Conference review	236
References	408,402	Data paper	5
Document contents		Editorial	3
Keywords plus (ID)	22,681	Erratum	27
Author's keywords (DE)	12,527	Letter	2
Authors		Note	9
Authors	15,996	Retracted	23
Authors of single-authored docs	238	Review	402
Authors collaboration		Short survey	3
Single-authored docs	360		
Co-authors per doc	4.14		
International co-authorships (%)	23.69		

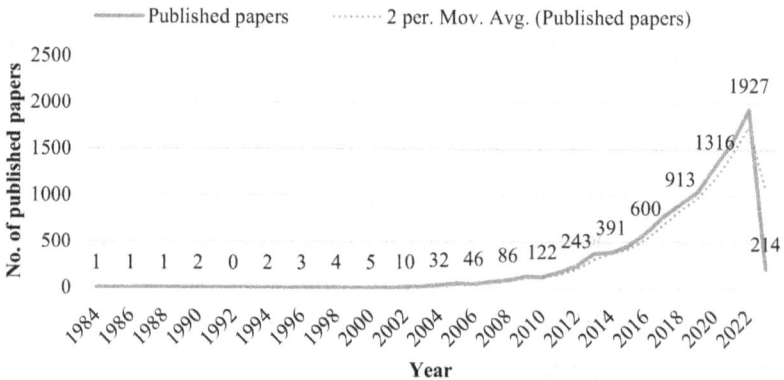

FIGURE 1.1   The evolution of annual publication of articles in the field of geopolymers.

1.2.2   RELATIONSHIP BETWEEN PUBLISHERS, AFFILIATED COUNTRIES, AND KEYWORDS

A three-field plot (Sankey diagram) of the most important top 10 numbers of items for the fields of country, sources, and keywords from the extracted data was created to depict the proportion of research topics for each country and the most targeted

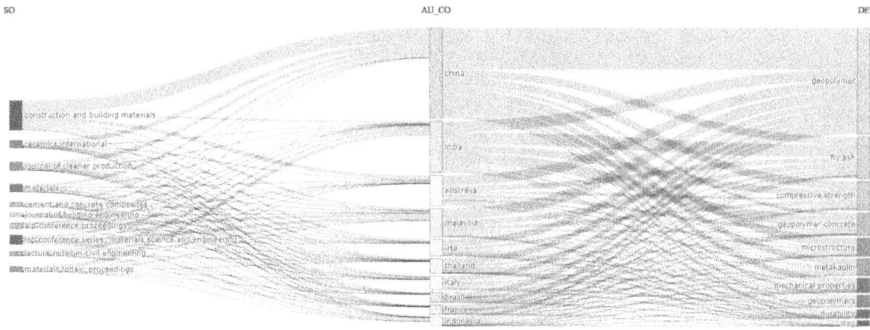

FIGURE 1.2 A three-field plot (Sankey diagram) of sources of publications (SO), authors' affiliations by countries (AU_CO), and keywords (DE) of the extracted data for the 10 most relevant topics.

journals by the authors in the geopolymers field. As presented in Figure 1.2, a very large part of the publication (almost 1,500) is associated with a researcher from China who published in the journal *Construction and Building Materials.* Moreover, China seems to be the country with the highest number of publications in the field that use "geopolymer" as a keyword. However, for the keyword "geopolymer concrete," the strongest link is presented for the researchers associated with India, while the most preferred publisher for these authors is Materials Today: Proceedings. As confirmed by the specifics of the identified journals, the publications on this topic are mostly on the designing, obtaining, and characterization of materials, especially construction materials. The main fields of application for these materials are also confirmed by the most commonly identified keywords, which are related to concrete or its main properties, i.e., compressive strength, durability, and microstructure.

1.2.3 THE JOURNALS WITH THE HIGHEST NUMBER OF PAPERS PUBLISHED IN THE GEOPOLYMERS FIELD

To further understand the trend on this topic, a chart with the top 25 most relevant journals was created. As shown in Figure 1.3, the journal *Construction and Building Materials* has more than 1,100 papers already indexed in Scopus as of the date of the analysis. There is no doubt that this publisher is by far the most wanted publisher in the world, with a total that is over three times greater than the total of the publisher ranked second and almost four times greater than the total of the publisher ranked third. Except for the first-ranked journal, the differences in the number of associated papers for the other 24 publishers are small (up to 20%), almost insignificant (0.1%), or nonexistent (e.g., the journals *Cement and Concrete Research* and *Ceramic Engineering and Science Proceedings*). As presented previously, most of the journals are related to materials science and engineering, but it is absolutely no surprise for the researcher from the geopolymers field to see the *Journal of Hazardous Materials* among them. Even though this journal is focused on research that can help us better comprehend the dangers that hazardous materials pose to human health and the

JOURNAL OF HAZARDOUS MATERIALS   78
JOURNAL OF THE AMERICAN CERAMIC..   80
JOURNAL OF PHYSICS: CONFERENCE SERIES   80
POLYMERS   89
MATEC WEB OF CONFERENCES   94
MATERIALS LETTERS   95
ADVANCED MATERIALS RESEARCH   95
CERAMIC ENGINEERING AND SCIENCE..   96
CEMENT AND CONCRETE RESEARCH   96
IOP CONFERENCE SERIES: EARTH AND..   101
APPLIED CLAY SCIENCE   112
CASE STUDIES IN CONSTRUCTION MATERIALS   116
JOURNAL OF MATERIALS IN CIVIL..   144
MATERIALS SCIENCE FORUM   151
KEY ENGINEERING MATERIALS   151
JOURNAL OF BUILDING ENGINEERING   158
AIP CONFERENCE PROCEEDINGS   171
CEMENT AND CONCRETE COMPOSITES   182
LECTURE NOTES IN CIVIL ENGINEERING   216
MATERIALS TODAY: PROCEEDINGS   226
JOURNAL OF CLEANER PRODUCTION   259
CERAMICS INTERNATIONAL   263
MATERIALS   **291**
IOP CONFERENCE SERIES: MATERIALS...   **352**
CONSTRUCTION AND BUILDING MATERIALS   1115

(Source) — No. of papers: 0, 200, 400, 600, 800, 1000, 1200

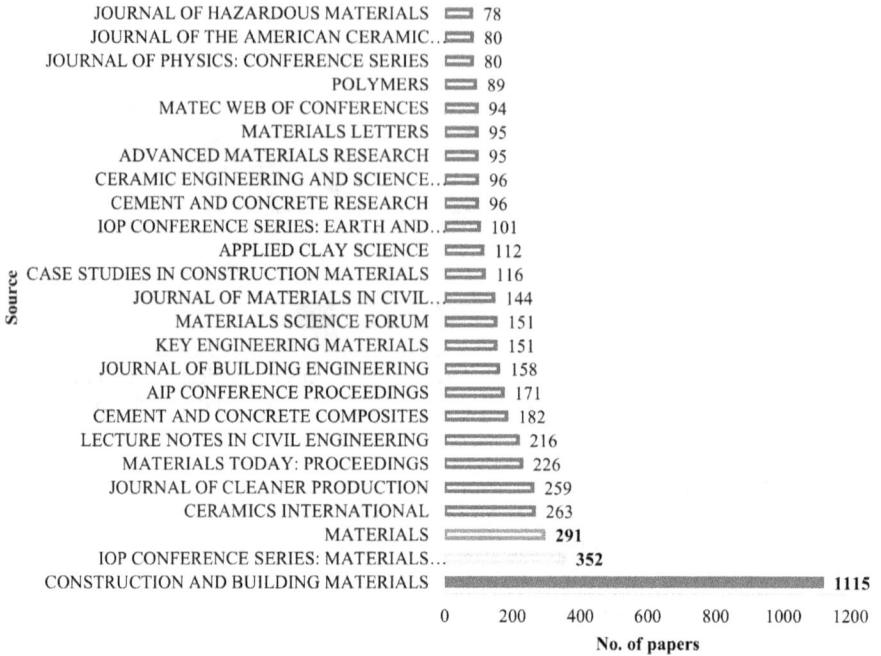

FIGURE 1.3 Top 25 most relevant sources with publications in the field of geopolymers.

environment, geopolymers fit its scope due to their advantages and applications in encapsulating dangerous wastes or eliminating soil contaminants.

By evaluating the sources' production over time and considering the top 10 journals, it was observed that the first paper was published by *Cement and Concrete Composites* in 2004. Surprisingly, in 2005, only the journal *Ceramics International* joined this field, while the leading journal, i.e., *Construction and Building Materials*, published its first paper on geopolymers in 2008. Also, starting in 2021, the number of publications indexed in Scopus for the IOP conference series on *Materials Science and Engineering* remained constant; a similar behavior can be observed for the line that corresponds to the journal *Ceramics International* (2018–2020), while the number of publications in the other journals steadily increased from year to year (Figure 1.4).

1.2.4 THE AUTHORS WITH THE HIGHEST IMPACT IN THE GEOPOLYMERS FIELD

When assessing an author's relevance in a certain sector, two factors should be considered: the author's production and the impact of those publications. Both of these indicators are revealed in Figure 1.5, which presents an overview of the top 25 most relevant authors in the field of geopolymers. The ranking is created considering the number of publications and the citations received each year. The number of papers published by an author in a certain period of time was used to assess productivity. In contrast, the influence was measured by the number of citations obtained each year. The analysis has been conducted to highlight the pioneers and current leaders in this

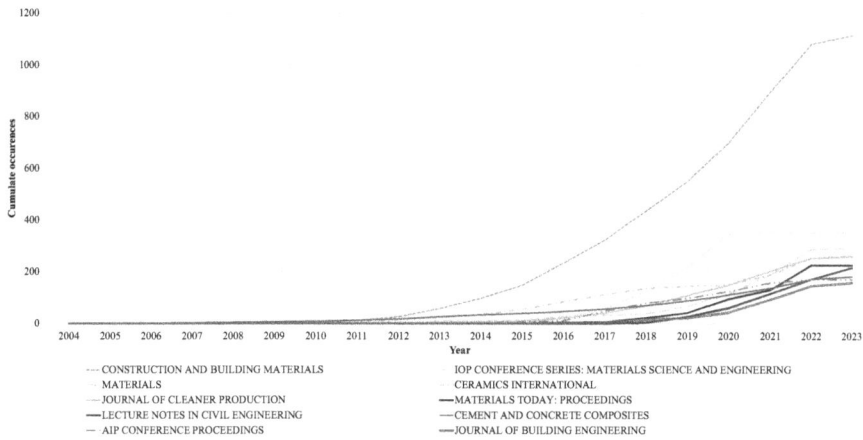

FIGURE 1.4 Cumulative distribution of the number of papers published by top 10 most relevant sources with publications in the field of geopolymers.

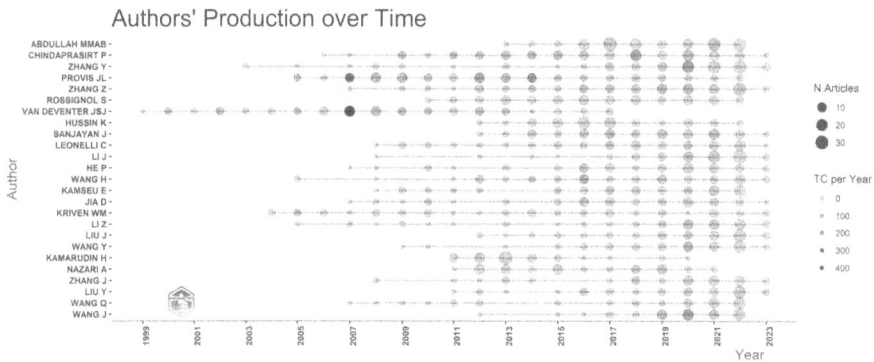

FIGURE 1.5 The production of top authors over time.

field. As can be seen from Figure 1.5, one of the editors of this book, i.e., Abdullah M.M.A.B., occupies the first position considering the authors' production over time, while the author with the longest uninterrupted line of articles, dating from 2004 to 2023, is Kriven W.M. Take note that the larger the circle, the more articles the author has written in that year. The deeper the circle, the more the citations obtained each year. To better understand the influence of these researchers in the past 10 years, the data provided by the software have been processed in Table 1.2 to identify the authors with the highest frequency of publication (freq), number of citations (TC), or production over time (TCpY). Accordingly, the highest value for each criterion from each year has been highlighted in bold. For example, in 2022, the highest frequency is associated with Li J, the highest TC is associated with Wang J, and the highest TCpY is associated with the same author as for TC. To make the data easier to follow, the authors that didn't obtain the highest score for one of the criteria during the evaluated period (such as He P., Jia D., Kamseu E., Kriven W.M., Leonelli C., Li Z., Rossignol S., Van Deventer J.S.J., Wan Q., Wang Y., Zhang J.) have been removed

TABLE 1.2

The Authors with the Highest Score in the Last 10 Years

Author	Year	2012	2013	2014	2015	2016	2017	2018	2019	2020	2021	2022	2023
Abdullah MMAB	Freq	–	2	3	3	20	34	22	10	14	28	28	–
	TC	–	70	33	128	194	508	176	32	222	279	68	–
	TCpY	–	6.36	3.30	14.22	24.25	72.57	29.33	6.40	55.50	93	34	–
Chindaprasirt P	Freq	7	11	8	10	11	9	19	10	18	12	12	2
	TC	913	921	1,057	892	479	288	959	173	420	114	34	0
	TCpY	76.08	83.72	105.7	99.11	59.87	41.14	159.83	34.60	105	38	17	0
Kamarudin H	Freq	16	35	16	20	29	19	6	4	10	5	2	0
	TC	1,327	622	392	313	513	312	40	33	122	26	1	0
	TCpY	86.16	56.54	39.20	34.77	64.12	44.57	6.66	6.06	30.50	8.66	0.50	0
Li J	Freq	1	–	1	–	1	2	6	7	14	22	30	5
	TC	2	–	1	–	21	0	232	187	283	201	60	1
	TCpY	0.16	–	0.10	–	2.33	0	38.66	37.40	70.75	67	30	1
Liu J	Freq	2	1	2	3	3	3	3	8	9	10	28	5
	TC	6	4	8	74	13	58	43	189	228	102	42	1
	TCpY	0.50	0.36	0.80	8.22	1.62	8.28	7.16	37.80	57	34	21	1
Liu Y	Freq	3	1	–	1	6	6	3	5	9	9	25	3
	TC	14	7	–	34	394	69	57	159	260	45	103	1
	TCpY	1.16	0.63	–	3.77	49.25	9.85	9.50	31.80	65	15	51.50	1
Nazari A	Freq	14	11	4	12	2	3	8	12	3	3	–	–
	TC	316	194	65	379	169	96	399	300	34	0	–	–
	TCpY	26.33	17.63	6.50	42.11	21.12	13.71	66.50	60	8.50	0	–	–

(Continued)

TABLE 1.2 (Continued)
The Authors with the Highest Score in the Last 10 Years

Author	Year	2012	2013	2014	2015	2016	2017	2018	2019	2020	2021	2022	2023
Provis JL	Freq	13	10	9	4	4	4	7	4	4	4	4	—
	TC	1,907	1,188	2,319	511	254	170	211	239	131	23	31	—
	TCpY	158.91	108	231.90	56.77	31.75	24.28	35.16	47.80	32.75	7.66	15.50	—
Sanjayan J	Freq	2	2	8	4	9	10	7	13	12	15	11	2
	TC	29	86	574	385	559	716	322	344	205	299	22	0
	TCpY	2.41	7.81	57.40	42.77	69.87	102.28	53.66	68.80	51.25	99.66	11	0
Wang H	Freq	7	2	3	4	11	4	5	9	6	7	17	3
	TC	657	140	771	380	1,167	197	196	424	47	55	41	0
	TCpY	54.75	12.72	77.10	42.22	145.87	28.14	32.67	84.80	11.75	18.33	20.50	0
Wang J	Freq	1	5	—	1	1	5	3	12	17	11	19	—
	TC	5	—	—	17	39	40	87	615	720	197	108	—
	TCpY	0.417	—	—	1.89	4.87	5.71	14.5	123	180	65.66	54	7
Zhang Y	Freq	—	4	—	11	2	10	7	10	21	26	27	0
	TC	—	164	—	11	61	376	245	321	792	167	67	0
	TCpY	—	14.91	—	1.22	7.62	53.71	40.83	64.20	198	55.66	33.50	0
Zhang Z	Freq	8	3	4	2	9	8	9	12	14	13	28	2
	TC	689	160	811	414	600	459	284	543	505	179	102	0
	TCpY	57.41	14.54	81.10	46	75	65.57	47.33	108.60	126.25	59.66	51	0

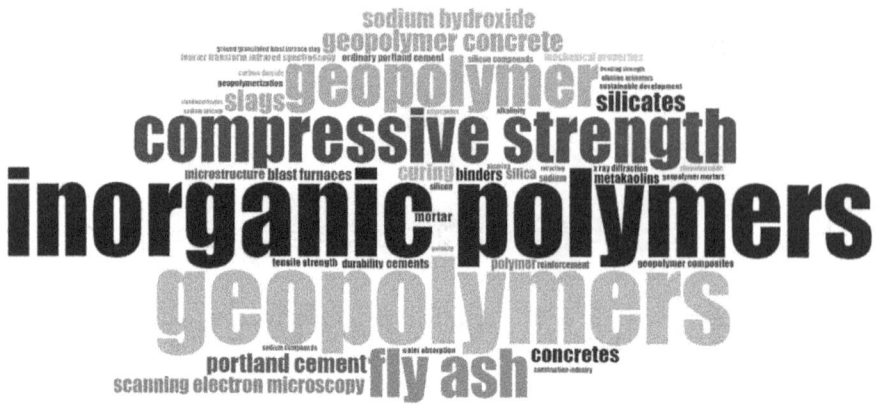

FIGURE 1.6   Word cloud illustration of top 50 keywords used by researchers in the retrieved data.

TABLE 1.3

The Most 25 Cited Papers in the Field of Geopolymers, According to Scopus Database

Position	Terms	Frequency	Position	Terms	Frequency
1	Inorganic polymers	7,129	6	Geopolymer concrete	1,794
2	Geopolymers	7,007	7	Slags	1,748
3	Compressive strength	4,402	8	Silicates	1,634
4	Geopolymer	4,182	9	Sodium hydroxide	1,548
5	Fly ash	3,938	10	Portland cement	1,534

from the table. Considering the mentioned parameters, it can be seen that Abdullah M.M.A.B. has the highest number of frequencies in 2017, 2018, and 2021, respectively. Provis J.L. had three consecutive years, i.e., 2012, 2013, and 2014, with the highest score for TC and TCpY.

The word cloud illustration (Figure 1.6) and Table 1.3 offer a visual representation of the frequency of the keywords used in the retrieved papers. It shows that the term "inorganic polymers" was the most frequently used keyword in the literature, indicating that it was the focus of the research. The other terms indicate the various components and properties of the geopolymers that were studied, as well as their main competitor, i.e., OPC.

Considering the worldwide distribution of researcher affiliations, Figure 1.7 shows that preoccupations in this field are prevalent on all continents and in almost any country. As can be seen, the countries that are missing from the map have no affiliated publications, while those colored in gray confirm that they have at least one paper indexed in Scopus. The intensity of the gray color increases depending on the number of published papers, i.e., it gets darker if the number of papers is higher. Also, the lines indicate at least ten jointly published papers between the indicated countries.

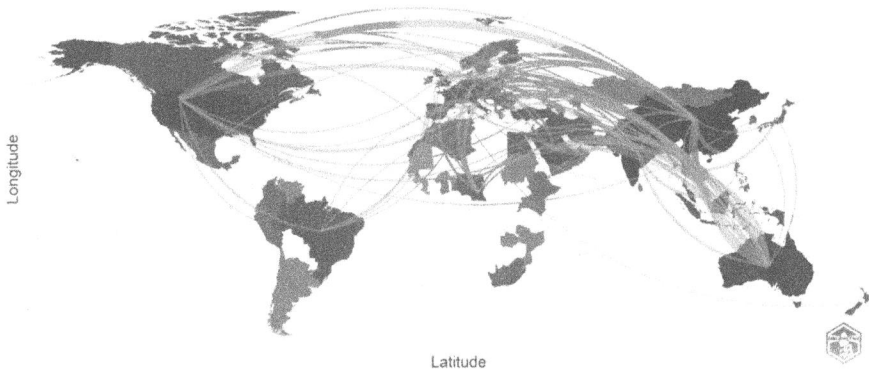

FIGURE 1.7 Country collaboration map of researchers in geopolymer technology.

Surprisingly, there is no paper affiliated with Russia, while the most intensive blue can be seen for China, India, Australia, and Malaysia, also confirmed by Figure 1.2.

1.2.5 THE PAPERS WITH THE HIGHEST IMPACT IN THE GEOPOLYMERS FIELD

The most cited paper in the field of geopolymers is a review type of paper, published by Duxson et al. (2007a) in 2007 (Table 1.4). The paper summarizes the previous research and briefly describes how the key parameters influence the final properties of the geopolymers. Accordingly, the authors focus their attention on how the processing condition and the characteristics of the activator influence the early properties (setting time, workability, etc.) and the structure of this product in order to define the obtaining technology. Following, as the second-ranked paper is Xu and van Deventer (2000), which deeply describes the geopolymerization reaction and the particularities of NaOH vs. KOH activation. Also, the authors show how the particularities of the raw materials, especially the content of Al-Si minerals, influence the compressive strength of geopolymer concrete by revealing the dissolution mechanism and ion-pair theory. In Duxson et al. (2007c), the concept of "green concrete" and the idea of replacing OPC-based materials with geopolymers toward a sustainable future were developed. During the literature review, the authors identified one limitation that is very strong even today, namely the lack of collaboration between academia and industry, which keeps the geopolymers out of wide industrial applications even today. Moreover, 2007 remains one of the years when research had to choose between developing geopolymers for ceramic applications or as substitutes for OPC-based materials. As the literature shows later, the construction materials sector becomes the most suitable one. In Duxson et al. (2005), a similar group of researchers conducted an experimental study to evaluate the relationship between the composition of the geopolymer and its mechanical and microstructural properties. The metakaolin-based geopolymer was activated with an alkaline solution, considering different ratios between Si and Al (1.15–2.15) to observe porosity and Young's modulus changes. In terms of porosity, the authors concluded that a ratio lower than 1.4 will result in a highly porous matrix (large pores), while the ratios above 1.65 will exhibit better homogeneity. The higher compressive strength was obtained at

TABLE 1.4

The Most 25 Cited Papers in the Field of Geopolymers, According to Scopus Database

Author	Year	Publishing Journal	References	Total Citations	TC per Year	Type of Paper
Duxson P	2007	*J. Mater. Sci-A.*	Duxson et al. (2007a)	2,838	166.94	Review
Xu H	2000	*Int. J. Miner. Process.*	Xu and van Deventer (2000)	1,293	53.88	Article
Duxson P	2007	*Cem. Concr. Res.*	Duxson et al. (2007c)	1,238	72.82	Review
Duxson P	2005	*Colloids. Surf. A. Physicochem. Eng. Asp.*	Duxson et al. (2005)	1,192	62.74	Article
Turner LK	2013	*Constr. Build. Mater.*	Turner and Collins (2013)	1,028	93.45	Article
McLellan BC	2011	*J. Clean. Prod.*	McLellan et al. (2011)	987	75.92	Article
Khale D	2007	*J. Mater. Sci.*	Khale and Chaudhary (2007)	951	55.94	Review
Hardjito D	2004	*Aci. Mater. J.*	Hardjito et al. (2004)	951	47.55	Article
Provis JL	2009	*Geopolymers: Struct. Process. Prop. and. Ind. Appl-A.*	Provis (2007)	882	58.80	Review
Singh B	2015	*Constr. Build. Mater.*	Singh et al. (2015)	767	85.22	Review
Nath P	2014	*Constr. Build. Mater.*	Nath and Sarker (2014)	741	74.10	Article
Habert G	2011	*J. Clean. Prod.*	Habert et al. (2011)	727	55.92	Review
Bakharev T	2005	*Cem. Concr. Res.*	Bakharev (2005a)	694	36.53	Article
Bakharev T	2005	*Cem. Concr. Res-A.*	Bakharev (2005b)	674	35.47	Article
Somna K	2011	Fuel	Somna et al. (2011)	665	51.15	Article
Van Jaarsveld JGS	2002	*Chem. Eng. J.*	van Jaarsveld et al. (2002)	656	29.82	Article
Komnitsas K	2007	*Minerals. Eng.*	Komnitsas and Zaharaki (2007)	647	38.06	Review

(Continued)

TABLE 1.4 (*Continued*)
The Most 25 Cited Papers in the Field of Geopolymers, According to Scopus Database

Author	Year	Publishing Journal	References	Total Citations	TC per Year	Type of Paper
Cheng TW	2003	*Minerals. Eng.*	Cheng and Chiu (2003)	647	30.81	Article
Ismail I	2014	*Cem. Concr. Compos.*	Ismail et al. (2014)	618	61.80	Article
Rattanasak U	2009	*Minerals. Eng.*	Rattanasak and Chindaprasirt (2009)	618	41.20	Article
Duxson P	2007	*Colloids. Surf. A. Physicochem. Eng. Asp.*	Duxson et al. (2007b)	616	36.24	Article
Chindaprasirt P	2007	*Cem. Concr. Compos.*	Chindaprasirt et al. (2007)	609	35.82	Article
Rovnaník P	2010	*Constr. Build. Mater.*	Rovnaník (2010)	601	42.93	Article
Luukkonen T	2018	*Cem. Concr. Res.*	Luukkonen et al. (2018)	594	99.00	Review
Kong DLY	2010	*Cem. Concr. Res.*	Kong and Sanjayan (2010)	591	42.21	Article

Si/Al = 1.9, assuming that higher ratios will conduct to multiple unreacted zones, while lower ratios will decrease the mechanical performance due to high porosity. Turner and Collins (2013) discussed one of the most important aspects of geopolymers' rapid development, namely their lower CO_2 emissions than OPC-based materials. The study quantifies the carbon dioxide equivalent emissions (CO_2-e) generated by all activities required to create one cubic meter of geopolymer or conventional concrete. The CO_2 footprint of geopolymers, calculated by the authors, was only 9% lower than that of OPC-based materials. The value was much lower than that published previously because, in this research, multiple parameters, such as raw material treatments, curing temperature, and alkaline activator manufacturing, were also considered. Surprisingly, even though the most compelling reason almost disappeared in 2013, the researchers continued to develop this field in order to develop materials with a lower CO_2 footprint by reducing the curing temperature, discovering cheaper activators, and involving raw materials without preliminary processing (sifting, drying, grinding, etc.). Two years earlier, McLellan et al. (2011) conducted research on the differences between the costs and CO_2 emissions of geopolymers compared to OPC. The study shows that geopolymers exhibit 44%–64% lower carbon emissions at a double price (especially because of sodium silicate).

In general, it can be seen that many studies are review papers (8 out of 25), but it is also important to note that among the most popular papers, there are some studies that show the CO_2 emissions or the activation mechanism of geopolymers that are of particular interest. These studies show that some of the most common uses for geopolymers include fire- and heat-resistant coatings and adhesives, medical applications, high-temperature ceramics, new binders for fire-resistant fiber composites, and new cements for concrete. Later (Luhar et al. 2023), due to their strong mechanical properties once solidified, even from recycled waste, along with their low negative effects on the environment, geopolymers attracted researchers interested in hazardous material encapsulation. Therefore, the literature shows that geopolymers have unique properties that provide an effective and economical means of safely and securely containing hazardous materials.

The present study conducts a bibliometric analysis in the field of geopolymers using Bibliometrix software, which provides the opportunity to quantitatively analyze the published literature. This technique was used to evaluate the studies indexed in the Scopus database to highlight the current trends in this field and the researchers that significantly contribute to the development of geopolymers.

The bibliometric analysis was performed on a significant set of data that included more than 10,000 published papers.

1.3 CONCLUSIONS

Although this work adds to the literature by emphasizing developments in the field of geopolymers, the bibliometric analysis has several limitations. These restrictions are attributable to Scopus indexing and the authors' inhomogeneity in describing the geopolymers. The database's limitations are connected to the gap between the publication date and the indexing date of the published literature. Furthermore, many articles are not indexed in the Scopus database. The constraints connected with heterogeneity in the literature are related to the various words used by researchers to substitute the term geopolymer, such as "green cement," "geocement," "eco-cement," and so on. Given these constraints, the study may not correctly capture the current status of the literature on the topic of geopolymers. However, its significance is demonstrated by the large volume of studied publications and the high quality of the database.

Despite the fact that geopolymers possess superior properties and, at the same time, can be obtained by simple methods, the largest amount of them is obtained from natural minerals. Therefore, it is essential to design and develop geopolymers that use mineral waste as a source of raw material, especially indigenous waste.

REFERENCES

Ahmad, J., A. Manan, A. Ali, M.W. Khan, M. Asim, and O. Zaid. 2020. A Study on Mechanical and Durability Aspects of Concrete Modified with Steel Fibers (SFS). *Civil Engineering and Architecture* 8, no. 5: 814–823.

Albino, V., A. Balice, and R.M. Dangelico. 2009. Environmental Strategies and Green Product Development: An Overview on Sustainability-Driven Companies. *Business Strategy and the Environment* 18, no. 2: 83–96.

Aria, M. and C. Cuccurullo. 2017. Bibliometrix: An R-Tool for Comprehensive Science Mapping Analysis. *Journal of Informetrics* 11, no. 4: 959–975.

Bakharev, T. 2005a. Resistance of Geopolymer Materials to Acid Attack. *Cement and Concrete Research* 35, no. 4: 658–670.

Bakharev, T. 2005b. Geopolymeric Materials Prepared Using Class F Fly Ash and Elevated Temperature Curing. *Cement and Concrete Research* 35, no. 6: 1224–1232.

Burduhos Nergis, D.D., M.M.A.B. Abdullah, P. Vizureanu, and M.F. Mohd Tahir. 2018. Geopolymers and Their Uses: Review. *IOP Conference Series: Materials Science and Engineering* 374, no. 1: 012019.

Cheng, T.W. and J.P. Chiu. 2003. Fire-Resistant Geopolymer Produce by Granulated Blast Furnace Slag. *Minerals Engineering* 16, no. 3: 205–210.

Chindaprasirt, P., T. Chareerat, and V. Sirivivatnanon. 2007. Workability and Strength of Coarse High Calcium Fly Ash Geopolymer. *Cement and Concrete Composites* 29, no. 3: 224–229.

Davidovits, J. 2015. *False-CO$_{2\text{-}Values}$* (Technical Paper #24, Scientific Papers). Geopolymer Institute Library, www.Geopolymer.Org. https://www.materialstoday.com/polymers-soft-materials/features/environmental-implications-of-geopolymers/,.

Dong, L., X. Tong, X. Li, J. Zhou, S. Wang, and B. Liu. 2019. Some Developments and New Insights of Environmental Problems and Deep Mining Strategy for Cleaner Production in Mines. *Journal of Cleaner Production*. doi:10.1016/j.jclepro.2018.10.291.

Duxson, P., J.L. Provis, G.C. Lukey, S.W. Mallicoat, W.M. Kriven, and J.S.J. van Deventer. 2005. Understanding the Relationship between Geopolymer Composition, Microstructure and Mechanical Properties. *Colloids and Surfaces A: Physicochemical and Engineering Aspects* 269, no. 1–3: 47–58.

Duxson, P., A. Fernández-Jiménez, J.L. Provis, G.C. Lukey, A. Palomo, and J.S.J. van Deventer. 2007a. Geopolymer Technology: The Current State of the Art. *Journal of Materials Science* 42, no. 9: 2917–2933.

Duxson, P., S.W. Mallicoat, G.C. Lukey, W.M. Kriven, and J.S.J. van Deventer. 2007b. The Effect of Alkali and Si/Al Ratio on the Development of Mechanical Properties of Metakaolin-Based Geopolymers. *Colloids and Surfaces A: Physicochemical and Engineering Aspects* 292, no. 1: 8–20.

Duxson, P., J.L. Provis, G.C. Lukey, and J.S.J. van Deventer. 2007c. The Role of Inorganic Polymer Technology in the Development of "Green Concrete." *Cement and Concrete Research* 37, no. 12: 1590–1597.

Geissdoerfer, M., P. Savaget, N.M.P. Bocken, and E.J. Hultink. 2017. The Circular Economy: A New Sustainability Paradigm? *Journal of Cleaner Production* 143: 757–768. https://api.semanticscholar.org/CorpusID:157449142.

Habert, G., J.B. D'Espinose De Lacaillerie, and N. Roussel. 2011. An Environmental Evaluation of Geopolymer Based Concrete Production: Reviewing Current Research Trends. *Journal of Cleaner Production* 19, no. 11: 1229–1238.

Hao, D.L.C., R.A. Razak, M. Kheimi, Z. Yahya, M.M.A.B. Abdullah, D.D.B. Nergis, H. Fansuri, R. Ediati, R. Mohamed, and A. Abdullah. 2022. Artificial Lightweight Aggregates Made from Pozzolanic Material: A Review on the Method, Physical and Mechanical Properties, Thermal and Microstructure. *Materials* 15, no. 11: 3929.

Hardjito, D., S.E. Wallah, D.M.J. Sumajouw, and B.V. Rangan. 2004. On the Development of Fly Ash-Based Concrete. *Materials Journal* 101, no. 6: 467–472.

He, Z., X. Zhu, J. Wang, M. Mu, and Y. Wang. 2019. Comparison of CO$_2$ Emissions from OPC and Recycled Cement Production. *Construction and Building Materials* 211: 965–973.

Horan, C., M. Genedy, M. Juenger, and E. van Oort. 2022. Fly Ash-Based Geopolymers as Lower Carbon Footprint Alternatives to Portland Cement for Well Cementing Applications. *Energies* 15: 8819. https://www.mdpi.com/1996-1073/15/23/8819/htm.

Ismail, I., S.A. Bernal, J.L. Provis, R.S. Nicolas, S. Hamdan, and J.S.J. van Deventer. 2014. Modification of Phase Evolution in Alkali-Activated Blast Furnace Slag by the Incorporation of Fly Ash. *Cement and Concrete Composites* 45: 125–135.

Jat, D., R. Motiani, S. Dalal, and I. Thakar. 2023. Mechanical Properties of Geopolymer Concrete Reinforced with Various Fibers: A Review. 139–156. doi:10.1007/978-981-19-6297-4_11.

Khale, D. and R. Chaudhary. 2007. Mechanism of Geopolymerization and Factors Influencing Its Development: A Review. *Journal of Materials Science* 42, no. 3: 729–746.

Khalifeh, M., A. Saasen, H. Hodne, R. Godøy and T. Vrålstad. 2018. Geopolymers as an Alternative for Oil Well Cementing Applications: A Review of Advantages and Concerns. *Journal of Energy Resources Technology, Transactions of the ASME* 140: 9.

Komnitsas, K. and D. Zaharaki. 2007. Geopolymerisation: A Review and Prospects for the Minerals Industry. *Minerals Engineering* 20, no. 14: 1261–1277.

Kong, D.L.Y. and J.G. Sanjayan. 2010. Effect of Elevated Temperatures on Geopolymer Paste, Mortar and Concrete. *Cement and Concrete Research* 40, no. 2: 334–339.

Li, X., D. Qin, Y. Hu, W. Ahmad, A. Ahmad, F. Aslam, and P. Joyklad. 2022. A Systematic Review of Waste Materials in Cement-Based Composites for Construction Applications. *Journal of Building Engineering* 45: 103447.

Luhar, I., S. Luhar, M.M.A.B. Abdullah, A.V. Sandu, P. Vizureanu, R.A. Razak, D.D. Burduhos-Nergis, and T. Imjai. 2023. Solidification/Stabilization Technology for Radioactive Wastes Using Cement: An Appraisal. *Materials* 16, no. 3: 954. https://www.mdpi.com/1996-1944/16/3/954/htm.

Luukkonen, T., Z. Abdollahnejad, J. Yliniemi, P. Kinnunen, and M. Illikainen. 2018. One-Part Alkali-Activated Materials: A Review. *Cement and Concrete Research* 103: 21–34. doi:10.1016%2Fj.cemconres.2017.10.001.

McLellan, B.C., R.P. Williams, J. Lay, A. van Riessen, and G.D. Corder. 2011. Costs and Carbon Emissions for Geopolymer Pastes in Comparison to Ordinary Portland Cement. *Journal of Cleaner Production* 19, no. 9–10: 1080–1090.

Nath, P. and P.K. Sarker. 2014. Effect of GGBFS on Setting, Workability and Early Strength Properties of Fly Ash Geopolymer Concrete Cured in Ambient Condition. *Construction and Building Materials* 66: 163–171.

Provis, J.L. and J.S.J. van Deventer. 2007. *Geopolymers Structures, Processing, Properties and Industrial Applications* (Vol. 1). London: Woodhead Publishing Series.

Rattanasak, U. and P. Chindaprasirt. 2009. Influence of NaOH Solution on the Synthesis of Fly Ash Geopolymer. *Minerals Engineering* 22, no. 12: 1073–1078.

Rovnaník, P. 2010. Effect of Curing Temperature on the Development of Hard Structure of Metakaolin-Based Geopolymer. *Construction and Building Materials* 24, no. 7: 1176–1183.

Samuvel Raj, R., G.P. Arulraj, N. Anand, B. Kanagaraj, E. Lubloy, and M.Z. Naser. 2023. Nanomaterials in Geopolymer Composites: A Review. *Developments in the Built Environment* 13: 100114.

Shah, K.W., G.F. Huseien, and T. Xiong. 2020. Functional Nanomaterials and Their Applications toward Smart and Green Buildings. *New Materials in Civil Engineering*: 395–433. doi:10.1016/B978-0-12-818961-0.00011-9

Shamsaei, E., O. Bolt, F.B. de Souza, E. Benhelal, K. Sagoe-Crentsil, and J. Sanjayan. 2021. Pathways to Commercialisation for Brown Coal Fly Ash-Based Geopolymer Concrete in Australia. *Sustainability* 13, no. 8: 4350. https://www.mdpi.com/2071-1050/13/8/4350/htm.

Singh, B., G. Ishwarya, M. Gupta, and S.K. Bhattacharyya. 2015. Geopolymer Concrete: A Review of Some Recent Developments. *Construction and Building Materials* 85: 78–90.

Somna, K., C. Jaturapitakkul, P. Kajitvichyanukul, and P. Chindaprasirt. 2011. NaOH-Activated Ground Fly Ash Geopolymer Cured at Ambient Temperature. *Fuel* 90, no. 6: 2118–2124.

Söylev, T.A. and T. Özturan. 2014. Durability, Physical and Mechanical Properties of Fiber-Reinforced Concretes at Low-Volume Fraction. *Construction and Building Materials* 73: 67–75.

Turner, L.K. and F.G. Collins. 2013. Carbon Dioxide Equivalent (CO_2-e) Emissions: A Comparison between Geopolymer and OPC Cement Concrete. *Construction and Building Materials* 43: 125–130.

Ul Rehman, M., K. Rashid, E.U. Haq, M. Hussain, and N. Shehzad. 2020. Physico-Mechanical Performance and Durability of Artificial Lightweight Aggregates Synthesized by Cementing and Geopolymerization. *Construction and Building Materials* 21: 1583–1588.

van Jaarsveld, J.G.S., J.S.J. van Deventer, and G.C. Lukey. 2002. The Effect of Composition and Temperature on the Properties of Fly Ash- and Kaolinite-Based Geopolymers. *Chemical Engineering Journal* 89, no. 1–3: 63–73.

Vizureanu, P. and D.D. Burduhos Nergis. 2020. *Green Materials Obtained by Geopolymerization for a Sustainable Future* (Vol. 90, p. 105) Millersville, PA: Materials Research Foundations. https://books.google.ro/books?hl=ro&lr=&id=GccLEAAAQB AJ&oi=fnd&pg=PP4&ots=QEyq5WN5Z_&sig=SCSckNzhyW-q1BMJKbNt1q5dy7E &redir_esc=y#v=onepage&q&f=false.

Xu, H. and J.S.J. van Deventer. 2000. The Geopolymerisation of Alumino-Silicate Minerals. *International Journal of Mineral Processing* 59, no. 3: 247–266.

2 Geopolymers Overview

Andrei Victor Sandu,
Dumitru Doru Burduhos-Nergis,
and Petrica Vizureanu
Gheorghe Asachi Technical University

Mohd Mustafa Al Bakri Abdullah
and Liyana Jamaludin
Universiti Malaysia Perlis

2.1 INTRODUCTION

Geopolymers are inorganic materials made of aluminum and silicon oxides that have been chemically balanced with alkali or phosphorus elements. The name "geo" refers to materials that have a structure akin to natural rocks but are man-made, and "polymer" refers to their chemical structure, which mimics organic polymers. Geopolymerization, the reaction that forms them, may occur at both room and high temperatures. Geopolymer not only outperforms OPC in many applications, but it also has several additional advantages such as quick curing, high acid resistance, excellent adhesion to aggregates, immobilization of toxic and hazardous compounds, and significantly lower energy consumption and greenhouse gas emissions (Andrew 2018; He et al. 2019; Gartner and Hirao 2015). Geopolymer, such as OPC, has brittle behavior with poor tensile strength and is susceptible to cracking (Ranjbar and Zhang 2020; Çelik et al. 2022). Furthermore, because of the high cost of sodium silicate, its industrial development and use are severely limited (Abdollahnejad et al. 2017). These shortcomings not only restrict structural design but also affect the long-term durability of structures. Geopolymer reinforcement using steel or carbon fiber has been researched (Zhang et al. 2016; Yan et al. 2016). Although these fibers may successfully improve geopolymer tensile strength, ductility, and toughness, they are all produced in an energy-intensive manner, and there is concern regarding how to dispose of them at the end of their life cycle (Isa et al. 2022). Growing environmental consciousness and the need to preserve the long-term viability of building materials have prompted initiatives to seek out alternative fibers.

Nonetheless, the vast majority of geopolymers now being made and researched are based on natural basic materials (kaolin). Metakaolin is preferred because of its quick rate of dissolution in the reactant solution, simple control of the Si/Al ratio, and white hue. However, its costly cost prevents it from being widely used in geopolymer composites (Mehmood et al. 2022). Several studies have shown that geopolymers with high performance may also be generated by using secondary source materials (industrial wastes such as fly ash or slag) (Cong and Cheng 2021). Generally, any amorphous material high in silicon and aluminum may be used to make geopolymers. This

DOI: 10.1201/9781003390190-2

explains why countries experiencing rapid industrialization are exhibiting a strong interest in this technology. These countries produce a significant volume of industrial waste and lack a well-defined recycling strategy. The use of garbage for geopolymer synthesis might not only reduce waste but also reduce basic raw material consumption.

Currently, the characteristics and final properties of geopolymers are directly dependent on the raw materials used (aluminosilicate source and activator). At the center of these interactions is the structure formed as a result of the geopolymer-ization chemical process, which occurs after merging the solid and liquid components. According to previous studies (Skariah Thomas et al. 2022; Abbas et al. 2022; Qaidi et al. 2022; Abdila et al. 2022), until recently, aluminosilicate sources such as metakaolin, coal ash, slag, mine tailings, and other types of mineral waste were used to produce geopolymers with properties similar to Portland cement-based products. They were both activated at the same time, with solutions having varied amounts of alkaline chemical components. In different combinations and concentrations, the most popular activators are based on Na_2SiO_3 and/or NaOH.

In recent years, however, researchers in the field have shifted to acidic solution acti-vation, utilizing H_3PO_4 in various chemical element ratios (Al/P and Si/P) between the base material and the activator (Zribi and Baklouti 2021). Regardless of the acti-vation mechanism or raw material used, the reaction results in a mineral polymer, bridging the gap between ceramic materials and organic polymer research. Polymer equivalence is achieved by including Si–O–Si (siloxo) groups in geopolymers with chemical structures comparable to polyethylene. A geopolymer is also defined as an extraordinarily long reticular string composed of silicon groups (SiO_4) and a specific tetragonal network of aluminum oxide (AlO_4) formed during the exothermic process of certain oligomers. Alkaline ions such as K^+, Na^+, or Li^+ balance the bonds between these tetrahedral structures (Vizureanu and Burduhos Nergis 2020; Joseph 2008).

The term was introduced in literature in the 1990s, by Davidovits (1991). Geopolymers, also known as inorganic polymers or binders activated, in most cases, by alkali metals, have generated interest globally. In general, geopolymers show amorphous or semi-crystalline structures and three-dimensional compounds of sili-con oxides. Geopolymers can be produced by mixing reactive materials containing aluminosilicates, such as fly ash, kaolin or metakaolin, blast furnace slag (BFS), and dolomite, with highly concentrated (3–12 M) alkaline solutions. The alkaline solu-tions most commonly used to produce geopolymers are sodium hydroxide (NaOH), potassium hydroxide (KOH), sodium silicate, and potassium silicate (Bell et al. 2009).

Some researchers have also explored the use of industrial wastes such as red mud, rice husk ash, and sewage sludge in the production of geopolymers. The exact composition and properties of a geopolymer will depend on the specific materi-als used and the conditions under which it is made (Rowles and O'Connor 2003; Ahmad et al. 2022).

Considering the current context of climate change, it is necessary to reduce the consumption of raw materials and energy, respectively, the reuse of materials, and even waste (Vizureanu and Burduhos Nergis 2020). Thus, geopolymers can be used in various applications. The field of the construction industry is becoming the area with the greatest potential compared to other applications such as refractory waste treatment, bio-materials, plastics, automotive, and aerospace.

Thus, through geopolymerization, we can obtain materials similar to OPC concrete using aluminosilicate waste (Zhang et al. 2020b). Basically, an aluminosilicic material is dissolved in an alkaline solution, obtaining a tetragonal structure of Si–O–Al following the partial elimination of water. Therefore, a geopolymer is an oxide material based on aluminum and silicon groups, chemically balanced by atoms of Na^+, K^+, etc., which is formed following the geopolymerization reaction.

Due to the simplicity of the obtaining process, the applications of geopolymers are varied, ranging from thermal insulations, fire-resistant materials, refractory linings, construction materials, materials for the metallurgical industry, and decorative objects, to the encapsulation of radioactive and toxic waste, etc. (Joseph 2008; Jiang et al. 2020; Khalifeh et al. 2018; Taylor et al. 2015).

However, geopolymers stand out as ecological concretes with low energy consumption and a small CO_2 footprint (Vizureanu and Burduhos Nergis 2020; Joseph Davidovits 2015). Therefore, geopolymerization is an advantageous technique for obtaining ecological materials that have similar or superior properties to those of conventional materials but use waste as a source of raw material.

Globally, several wastes with geopolymerization potential have been identified, such as fly ash, red mud, BFS, etc. After the chemical reaction between the solid raw material (waste) and the alkaline activator (a solution of sodium silicate or potassium silicate and sodium hydroxide), an inorganic material with a structure similar to that of zeolites is obtained.

Thus, geopolymers are formed by a chemical reaction known as geopolymerization, which occurs as a result of mixing at least two constituents (a solid rich in silicon and aluminum oxides and an alkaline solution). Geopolymerization begins with the dissolution of silicon, aluminum, and calcium hydrates (if any) from the solid material (raw material) under the action of the alkaline activator. In the second stage, nucleation, oligomerization, polymerization, and polycondensation take place; therefore, groups of atoms are reoriented and groups called polysialates are created.

In the field of construction, geopolymers are generally used as a substitute for ordinary Portland cement (OPC)-based materials. In the manufacture of concrete, geopolymers are used as binders due to their exceptional thermal performance, superior mechanical properties, and considerable durability. The main motivating factor in the replacement of conventional concretes (those based on OPC) with geopolymer ones is related to the environmental problems regarding the enormous energy consumption and carbon dioxide (CO_2) emissions during their manufacture (Joseph Davidovits 2015; Papa et al. 2019). Industrial concrete production is one of the biggest contributors to global warming. The development of the cement industry is one of the real contributors to air pollutants. It has been calculated that 13,500 million tons of CO_2 are created from this process worldwide, accounting for about 7% of ozone-depleting substances delivered each year (Xue et al. 2023).

Nowadays, the construction industry has started to emphasize green technology and eco-friendly materials. Geopolymers have been introduced into the industry, offering virtually the same performance as regular cementitious pavements in various applications but with lower greenhouse emissions. Table 2.1 shows the comparison between geopolymers and regular concrete.

TABLE 2.1
Comparison between Geopolymers and Regular Concrete (Xue et al. 2023; Jat et al. 2023)

Conventional Concrete	Geopolymers
• Cement production is carried out in a separate place before proceeding to the mixing process	• The process of mixing all the ingredients is done simultaneously
• Produces carbon dioxide in high amounts	• Lower carbon dioxide production and more environmentally friendly
• Must be compacted and produce rough surfaces	• Contains adjustable properties (self-leveling) and produces a smooth surface
• Design method based on cement/water mix ratio	• Design method based on chemical ratio and pozzolanic materials

Geopolymers have achieved huge performance potential compared to the current standard of ordinary cement. Meanwhile, its processing can be done either at a lower temperature or at room temperature with lower carbon dioxide emissions. In most geopolymer production, 70% of the ultimate compressive strength can be achieved within the first 4 hours after curing (Wang et al. 2021).

2.2 GEOPOLYMERS CHEMISTRY

These oxidic materials are formed by means of a chemical reaction known as geo-polymerization, which occurs as a result of the mixing of at least two constituents (a solid rich in silicon and aluminum and an alkaline solution). Geopolymerization starts with a step of dissolving silicon, aluminum, and calcium hydrates from the solid material (raw material) under the action of the alkaline activator. In the second stage, nucleation, oligomerization (formation of oligomers), polymerization, and polycon-densation occur, and thus the groups of atoms reorient and form groups named, by Davidovits, polysialates (eng. polysialate), the term being a generic name of oxide groups of aluminosilicates (Table 2.2) (Davidovits 1991). In other words, after the dissociation of the Si–O–Si and Si–O–Al compounds, very reactive Al^{3+} and Si^{4+} species are released, which, following the interaction between them, form oligomers of SiO_4 and AlO_4 that are organized into 3D polymer chains of Si–O–Al–O eliminat-ing water (Provis 2009; Burduhos Nergis et al. 2018). The type and structure of the compounds in the polymer chains formed are determined by the ratio between Si and Al based on the empirical relationship (Equation 2.1) (Burduhos Nergis et al. 2018):

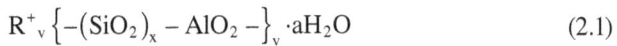

$$R^+_v\left\{-(SiO_2)_x - AlO_2 -\right\}_v \cdot aH_2O \qquad (2.1)$$

where R+—the alkaline cation of the activator (Na+, K+, etc.); v—degree of polym-erization; x—the ratio between Si/Al; and a—the number of water molecules (amount of water).

The ratio (x) between Si and Al can have values between 1 and 300, if the value of (x) is less than 3 (x < 3), then the geopolymer will exhibit high adhesion and flexibility properties due to the linear 2D structure. As the ratio (x) increases,

TABLE 2.2
Types of Polysialates Present in Geopolymers

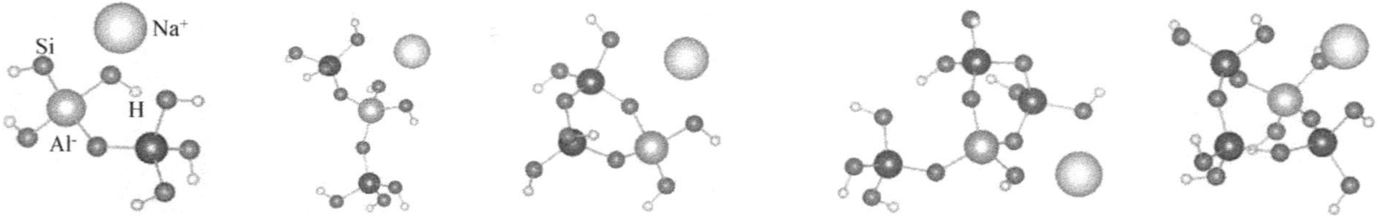

Na⁺ → Na^+

Si

H

Al⁻

(a)

(b)

(c)

(d)

(e)

Polysialate
Si/Al ≈ 1
(–Si–O–Al–O–)

Polysialate–siloxo
Si/Al ≤ 2
(–Si–O–Al–O–Si–O–)

Polysialate–disiloxo
Si/Al ≤ 3
(–Si–O–Al–O–Si–O–Si–O–)

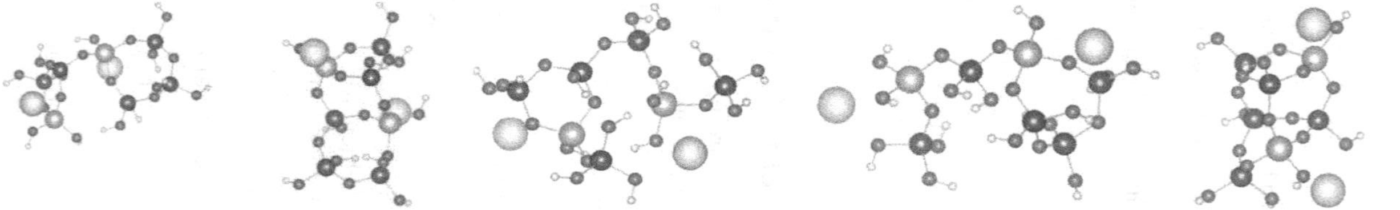

c+e;

b+d;

b+e;

c+e;

Source: Adapted with Permission from Kołeżyński et al. (2018).

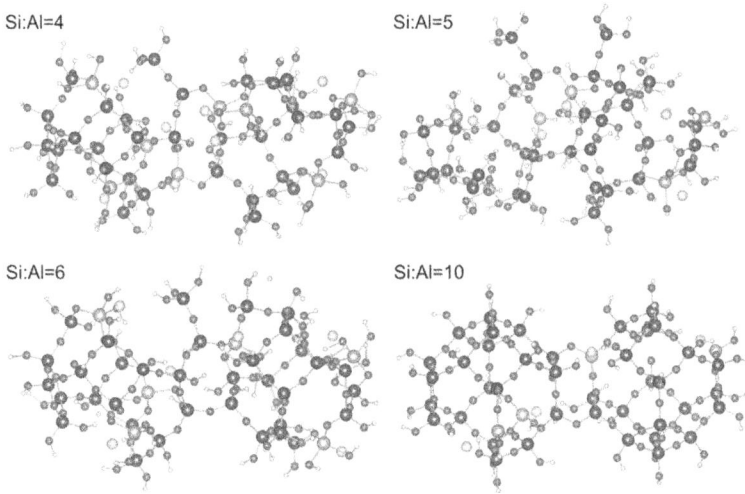

FIGURE 2.1 Arrangement of sialate bonds according to the ratio of Si to Al. Reprinted with permission from Koleżyński et al. (2018).

the fragility of the final structure increases, and its reticular network becomes 3D (Figure 2.1). At the same time, it was observed that with the increase in the ratio between silicon and aluminum oxide, there is also an increase in the curing period (Koleżyński et al. 2018).

Although the activator's chemical element is depicted as sodium ions in the preceding images, the literature indicates that it could be changed to potassium, lithium, calcium, or phosphorus. As a result, polysialates might have the chemical formula (Na, K, Li, Ca, P) (sialate unit) or (Na, K, Li, Ca, P) (sialate unit) (Si–O–Al). Furthermore, after the condensation of certain orthosialate monomers with $Si(OH)_4$ ortho-silicic acid groups, polycondensation of $(OH)_3$–Si–O–Al–$(OH)_3$ monomers leads to the formation of (–Si–O–Al–O–) or (–Si–O–Al–O–Si–O–). In the case of geopolymers with a high Fe content, a portion of the Si atoms can be replaced by Fe atoms, resulting in structures similar to the ones shown before. For example, in the initial phase of acid activation of metakaolin (a raw material rich in Al and Si), the ortho-sialate monomer is generated, followed by its condensation into cyclo-disialate, the precursor unit of hydrosodalite.

Thus, geopolymer refers to a three-dimensional material of amorphous to semi-crystalline silicon aluminate. In addition, geopolymer is formed by initiating aluminosilicate source materials, such as clay, fly ash, and slag with an alkaline solution. These materials exist naturally, but also as industrial waste or by-products of materials that require relatively less manufacturing energy. Exposure of these thermally activated aluminosilicates to the highly alkaline environment (combination of silicates and hydroxides) contributed to the formation of a two-dimensional Si–O–Al structure. Moreover, geopolymeric materials respond to a demand for cement-based products that confer zeolitic and ceramic properties (Elettra Papa et al. 2018).

Geopolymers arose through a series of complex reaction processes initiated by pozzolanic activation. The main processes were the dissolution of the aluminosilicate component in an alkaline medium, the polymerization of dissolved minerals in the gel, the precipitation of the formed hydration products, and the final strengthening of the matrix by excluding excess water (van Deventer et al. 2012).

Thus, the first reaction involved is dissolution, which occurs directly at the contact between the aluminosilicate and the alkaline solution, thus allocating the ionic interface and breaking the covalent bonds between silicon, aluminum, and oxygen atoms. Then, the polymerization process takes place, through which a rapid chemical reaction occurs in an alkaline environment on Si–O–Al bonds. The geopolymer gel formed contains alkaline cations that compensate for the deficient charges subjected to the substitution of aluminum with silicon (Xu and van Deventer 2000).

The reactions proceed with moderate development of crystal structure as the polymerized gel nuclei reach crucial size. The crystallinity of the matrix corresponds to the rate at which the precipitate forms. Spontaneous reactions between alkali and ash reduce the growth time of a crystalline medium, which is representative of typical zeolites (Rożek et al. 2019). However, the final product of geopolymers consists of an amorphous and semi-crystalline "cement" phase.

2.3   RAW MATERIALS

When we speak about raw materials, there are a huge number of materials that can be used for geopolymers, mainly those rich in aluminum and silicon oxides.

An aluminosilicate is described as a series of finely divided siliceous, aluminous, and siliceous materials that react chemically with slaked lime at ordinary temperature and in the presence of moisture to form slow-setting cement. Silica-rich material (fly ash or slag) and aluminum-rich material (clay) become essential prerequisites for geopolymerization.

2.3.1   CLAY

Kaolinite is one of the normal clay minerals used in the production of geopolymers. Kaolinite contains a 1:1 uncharged dioctahedral layer structure, as shown in Figure 2.2, with the synthetic formula $Al_2O_3 \cdot 2SiO_2 \cdot 2H_2O$ (Liew et al. 2016).

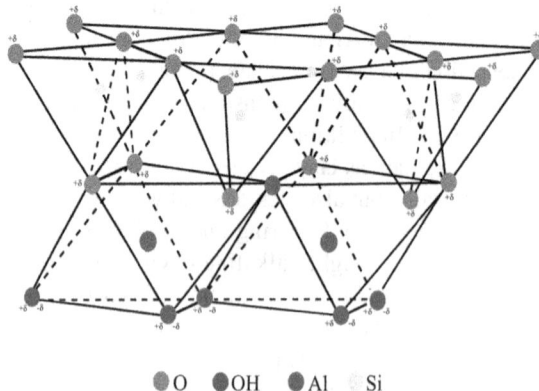

○ O   ● OH   ● Al   Si

FIGURE 2.2   Kaolinite structure (Liew et al. 2016).

The compounds $(Si_2O_5)n^{2-}$ and $Al(OH)_3$ are connected by sharing oxygen atoms that were held together by hydrogen and van der Waals bonds (Li et al. 2010a). Finally, kaolinite contains a small surface area for geopolymerization compared to fly ash, which has round particles. Thus, the small surface area allows only reduced dissolution by the alkaline reactant and consequently obtains a lower resistance (Heah et al. 2012). In addition, using only kaolinite in the geopolymerization reaction produced a weak structure.

Usually, better mechanical properties are evidenced by geopolymers generated from heat sources or calcination, such as BFS, fly ash, and metakaolin (Khale and Chaudhary 2007; Kanagaraj et al. 2023). The sintering process helps to improve the reactivity of kaolinite to the geopolymerization response. Therefore, favorable calcinations of kaolinite lead to strongly pozzolanic amorphous phases. These amorphous phases provide an active constituent that acquires the maximum strength of the geopolymer. At a given number of calcinations processed at 550°C–800°C, the strongly bound hydroxyl ions in the constitutive Al layer are dehydroxylated by water loss (Nikolov et al. 2020; Kenne Diffo et al. 2015). Thus, the transformation of kaolinite into the disordered metastable phase of metakaolin took place.

According to Rowles et al. (Rowles and O'Connor 2003), metakaolin (Figure 2.3) can be produced by heating kaolinite to 750°C in air for 24 h. Metakaolin can be generated at the same calcination temperature; however, it is done in a shorter time period of almost 10 hours. Above 900°C, metakaolin obtained low mechanical properties, probably caused by overheating during calcinations, resulting in non-reactivity toward crystalline phases.

FIGURE 2.3 Metakaolin image (Dai et al. 2022).

Clay is a naturally occurring mineral resource that is extensively spread. It is an aluminosilicate salt with minuscule (2 mm) particles. It is an earthy rock that is cohesive and flexible. Clay is a type of layered silicate made up of an alumina octahedral layer and a silicon oxygen tetrahedral layer. Clay minerals, including kaolin, zeolite, and others, are frequently utilized as precursors for the creation of geopolymers due to their compositional characteristics.

Dolomite, often referred to as kaolin, is a white, fine-grained, soft clay that has good flexibility and fire resistance. By dehydrating kaolin at the proper temperature, metakaolin (MK), an anhydrous aluminum silicate, is created (600°C). It displays the typical MK, which has been widely employed in the creation of geopolymers. The MK-based geopolymers have thermal insulating qualities, high compressive strength, strong bonding, etc. Many researchers have combined various substances with the system in order to lower costs, maintain high performance, and achieve resource reuse due to the superior mechanical qualities of MK-based geopolymers. Additionally, the study creates porous geopolymers based on kaolinite that reduce heat transfer while also being acoustically quiet. A type of aluminosilicate mineral called zeolite contains alkali metal or alkali earth metal and has properties including adsorbability, ion exchange, catalysis, acid resistance, and heat resistance. The geopolymer created by alkali-excited natural zeolite retains the porosity characteristics of zeolite while having the mechanical strength of geopolymer gel.

2.3.2 SLAG

An ironmaking by-product known as BFS can be produced at a temperature of about 1,500°C. Moreover, BFS is frequently referred to as slag. Because of its amorphous nature, high hardness, and pozzolanic activity, ground granulated BFS (GGBFS), also known as BFS cooled in water, is primarily utilized as a partial replacement for OPC after grinding. It is possible to get a decent reaction rate at a temperature as low as 0°C thanks to GGBS's high reactivity toward the synthesis of geopolymers. Less heat is produced during hydration when slag is used in place of cement, which lowers the likelihood of cracking.

GGBFS is one of the industrial by-products generated by the rapid water cooling of molten steel. This is generated following the production of first-fusion pig iron, as a result of the introduction of slag agents such as coke ash, iron ore, and limestone, which were mixed with the iron ore to remove impurities. During the iron ore reduction process, molten slag is formed as a non-metallic liquid (basically consisting of silicates, aluminates, calcium, and other bases) that floats on top of the molten iron (Li et al. 2010b). The molten slag and liquid metal are then separated before the cooling process. Different cooling methods lead to the production of different types of slag, namely expanded, air-cooled, and granulated.

GGBFS was produced from molten slag quenched sufficiently rapidly by water and classified as a hydraulic latent material containing cementitious and pozzolanic characteristics. In addition, GGBFS is commonly used as an additive material in ordinary cement concrete during cement hydration due to the improvement of its reactivity (Figure 2.4).

FIGURE 2.4 Granulated slag (Ziada et al. 2021).

The chemical composition of the slag is crucial to a balanced calcium aluminosilicate framework with overburden. The main component of GGBFS is the $CaO-SiO_2-MgO-Al_2O_3$ network, which is defined as a combination of depolymerized calcium aluminosilicate glass with similar akermanite ($2CaO \cdot MgO \cdot 2SiO_2$) and gehlenite ($2CaO \cdot Al_2O_3 \cdot SiO_2$) (Qureshi and Chish 2013). The key cations that form the glassy structure are Si^{4+} and Al^{3+}, and the divalent Ca^{2+} and Mg^{2+} act as network modifiers along with any alkali present (Provis 2007). The abundant presence of Ca^{2+} that coexisted in the glass structure of GGBFS showed better cement reactivity than the geopolymer material consisting of homogeneous glass and crystalline phases.

Granulated BFS has various industrial applications. Typically, GGBFS is used as a cement replacement (30%–50%) in normal concrete, while it can replace up to 70% in heavy-duty marine concrete applications. Moreover, GGBFS has been used most often in the cement industry. For example, GGBFS can be used separately or together with regular cement clinker and calcium sulfate. Underground GGBFS can be applied as a normal weight aggregate in concrete and is suitable as a base-course material in road construction. Some uses of GGBFS include glass manufacturing, sports field sub-base (drainage application), sandblasting media requiring fine etching, and concrete block manufacturing.

2.3.3 Ash

There is a constant expansion in the use of non-renewable energy sources to create vitality. Coal is mostly used to create steam in the modern and contemporary eras. Thus, coal ash is acquired as a waste that should be disposed of in a natural way. However, due to the huge amount created, most types of ash, including fly ash and bottom ash, are disposed of in landfills. This transfer process causes different problems.

FIGURE 2.5   Fly ash (Ziada et al. 2021).

The ash, generally called fly ash, is produced during combustion and contains fine particles that rise with the combustion gases. This is usually captured by electrostatic precipitators or other gas filtering equipment before it reaches the stacks of coal-fired power plants. Depending on the source of coal that is burned, the physical and chemical characteristics of fly ash change considerably. However, all fly ash contains significant concentrations of silicon dioxide (SiO_2), aluminum oxide (Al_2O_3), and calcium oxide (CaO) (Deb et al. 2014; Tho-In et al. 2016).

Fly ash (Figure 2.5) is generally classified into two types: one is class C (high calcium) and the other is class F (low calcium). Fly ash that has been obtained from sub-bituminous coals, according to ASTM C618 (2012), is class C, containing more than 20% calcium oxide (CaO). Meanwhile, class F, which includes a low calcium composition, comes from the consumption of bituminous coal and anthracite. Crucial parameters to be considered in the selection of fly ash as raw material are silicon dioxide, amorphous phase, grade, and, in addition, calcium content (Belviso 2018).

2.3.4   Dolomite

Calcium carbonate materials, for example, calcite ($CaCO_3$) and dolomite ($CaMg(CO_3)_2$), are naturally abundant and mostly economic minerals. Dolomite is one of the most recognizable carbonate minerals in geological structures. It is an anhydrous carbonate mineral created from calcium, magnesium, and carbonate, otherwise known as $CaMg(CO_3)_2$. The word dolomite is additionally used to describe sedimentary carbonate rock or rock that is made mostly of the mineral dolomite (otherwise called

FIGURE 2.6 Dolomite (Nowak et al. 2022).

dolostone). Dolomite is one of the natural resources that will be used as a feedstock for geopolymer composites. The mineral dolomite will solidify in a trigonal-rhombohedral framework. Its colors are dark white, brown, or crystal pink. Dolomite contains double carbonate, which has an alternate structural arrangement of calcium and magnesium particles. Unlike calcite, dolomite does not dissolve rapidly in a dilute corrosive medium (hydrochloric acid) (Sotelo-Piña et al. 2018; Zhang et al. 2020a).

Dolomite forms in an alternative class of crystals (Figure 2.6), ranging from the calcite group of minerals. This can be identified by the more elongated crystal shapes of dolomite compared to those formed by calcite. In addition, calcite group minerals occur in scalenohedral crystals, while dolomites never occur as such crystals (Zhang et al. 2020a; Yip et al. 2005).

2.3.5 LATERITE SOIL

Aluminosilicates, iron, and aluminum are abundant in the mineral laterite. Due to its excellent resistance to corrosion, the majority of laterite is reddish-brown and has been used for a long time as conventional brick, roads, and buildings. The manufacture of geopolymers has recently undergone a change based on lateritic, which has excellent mechanical strength (Mathew and Issac 2020; Subaer et al., 2019). Moreover, the raw material for the geopolymer type Na-poly, laterite, has an outstanding molar oxidation ratio ($SiO_2/(Al_2O_3 \ Fe_2O_3)$) (sialate-siloxo). The laterite geopolymer's microstructure and mechanical characteristics are greatly influenced by the molar oxide ratio of silica to alumina. Furthermore, laterite and other solid wastes are combined to create high-strength geopolymers.

In non-load-bearing building materials, laterite (Figure 2.7) and mixed laterite-slag geopolymer offer promising application prospects.

2.3.6 RED MUD

Red mud is a by-product of the industrial Bayer process used to refine aluminum (Figure 2.8). The Bayer method uses sodium hydroxide to dissolve the soluble portion of bauxite at high temperatures and pressures. The RM will invariably have

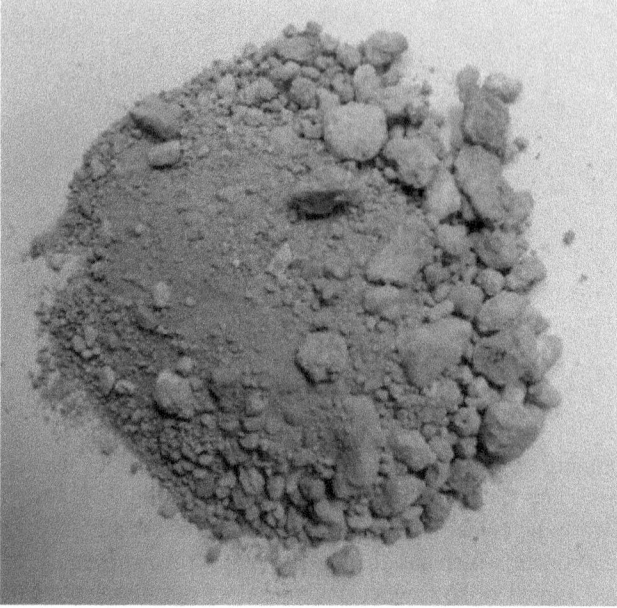

FIGURE 2.7   Laterite soil (Maichin et al. 2021).

FIGURE 2.8   Red mud (Nie et al. 2020).

a high pH since a tiny amount of the sodium hydroxide utilized in this process is left behind. By taking advantage of RM's high alkalinity, using it as mud reduces the overall amount of alkali activator, saving time and energy while also lowering the cost of the geopolymer's manufacturing. The ideal replacement value of RM for FA-based geopolymers varies depending on the NaOH concentration and curing circumstances.

2.3.7 OTHER RAW MATERIALS

The materials such as RHA as the primary biomass ash, fly ash, BFS, and RM all show high silica and alumina contents, which are acceptable for the additional materials as gelling materials. Steel slag (STS), silica fume (SF), volcanic ash (VA), waste glass (WG), coal gangue (CG), high-magnesium nickel slag (HMNS), and other minerals are also frequently employed. The amorphous structure and plentiful silicon and aluminum components found in the waste catalyst residue released from a variety of industrial goods can be exploited to create synthetic geopolymers with compressive strengths of up to 40–85 MPa.

2.4 FACTORS THAT INFLUENCE THE GEOPOLYMER PROPERTIES

Various industrial by-products offer different physical, chemical, and mineralogical characteristics, so the rate of geopolymer formation is influenced by different parameters such as curing method, alkali concentration, and mix design. In addition, the effects of major factors on the manufacture of geopolymers depend on the type of chemical activator used, the particle size distribution of the source material, the manufacture of the curing regime, and the exposure to the aggressive environment.

2.4.1 ALKALINITY

The concentration of MOH solution ($M = Na^+$, K^+, etc.) has a significant impact on the mechanical performance of geopolymers. Alkalis encourage the dissolution and solubility of aluminosilicates and also increase the rate of the geopolymerization reaction (Yao et al. 2009). The choice of an alkaline activating solution is an important criterion, and NaOH solution and KOH solution have been mainly used as suitable activating agents. However, NaOH solution was commonly preferred for its higher efficiency in releasing silicates and aluminates from the precursor material as well as economic effectiveness (Provis 2007).

Typically, increasing NaOH concentration in the range of 4–12 M increases the strength of metakaolin geopolymers. XRD analysis revealed that the amorphous phase develops with increasing NaOH concentration (Wang et al. 2005). During the dissolution reaction, NaOH provided OH^- and Na^+, which are like catalysts for the aluminosilica reaction.

Alkaline solutions used as activators in geopolymer technology can be classified as follows (R—represents the alkaline cation) (Joseph 2008):

- Alcaline, ROH;
- Low alkalinity salts, R_2CO_3, R_2SiO_3, R_3PO_4;
- Silicates, $R_2O \cdot nSiO_3$;
- Aluminates, $R_2O \cdot nAlO_3$;
- Aluminosilicates, $R_2O \cdot nAl_2SO_3 \cdot (2-6)SiO_2$.

The researchers concluded that a high concentration of alkali damages the strength of geopolymers. For example, the strength of geopolymers increased with NaOH concentration and decreased after an optimum concentration was reached. The optimum concentration of NaOH at 9 M was obtained for metakaolin geopolymers. Above this optimum point, the polymerization reaction is unacceptable.

2.4.2 SOLID-TO-LIQUID RATIO

During the production of geopolymers, the solid part is represented by aluminosilicates and the liquid part by the alkaline reactant. The solid-to-liquid ratio is crucial because it determines the optimal amount of solid and liquid for the mixing process, affecting workability, dissolution, geopolymerization reaction, and therefore the final strength of the product. The S/L ratio lower than 0.8 showed better workability during the geopolymerization reaction (Vizureanu and Burduhos Nergis 2020; Burduhos Nergis et al. 2018). The initial precursor solid content was affected by the formation rate of the geopolymer network; the kaolin-based geopolymer with the lowest S/L ratio (0.60) required more time to cure. Also, a weak structure was initially formed due to its low reactivity, which led to the low rate of development of geopolymerization (Komnitsas and Zaharaki 2007).

The solid-liquid ratio has an associated degree of particular impact on the workability of the geopolymer paste. A higher S/L ratio contributed to lower workability. Otherwise, the low S/L ratio slows down the geopolymerization reaction due to the inter-particle interaction of the precursor materials, which prevents the increase of geopolymer workability (Heah et al. 2012).

2.4.3 SODIUM SILICATE-TO-SODIUM HYDROXIDE RATIO

Alkaline activator ratios play a major role in the geopolymerization reaction. The ratio of NaOH solution to liquid Na_2SiO_3 was essential to forming the geopolymer. In fact, Na_2SiO_3 acts as a binder and NaOH acts as a solvent in the geopolymerization reaction. The Na_2SiO_3/NaOH ratio influenced the workability of the geopolymers. Workability has been identified as a property of fresh binder that measures the ease with which the paste can be mixed, placed, consolidated, and finished (Deb et al. 2014). The sodium silicate liquid generates a high viscosity compared to the NaOH solution, simultaneously improving the sodium silicate content and thus improving the workability of the geopolymer.

Based on the study by Sathonsaowaphak et al. (2009), the strength of geopolymer was developed with increasing the Na_2SiO_3/NaOH ratio to 1.5. Increasing the sodium silicate content in the alkaline activator solution affected the workability and setting times of the fly ash-based geopolymer. The decrease in workability and setting times of the geopolymer mixture was included in the list of difficulties in the formation of reinforced geopolymer. Therefore, the difficulty of consolidation led to a lower degree of geopolymerization and, implicitly, its strength.

Otherwise, the alkaline activator ratio should be considered in terms of the SiO_2/Na_2O molar ratio. Increasing the SiO_2/Na_2O ratio leads to a passive reaction and interrupts the setting of the paste. The system with Na silicate solution obtained a lower reaction rate than that with K silicate solution (Rahier et al. 2007). Economically, granule NaOH is more economical than liquid Na_2SiO_3, so it promotes the use of a lower ratio of alkaline activator in the geopolymerization reaction without demolishing the workability and strength of the cured product.

2.5 OBTAINING METHODS

The process of obtaining a geopolymer (Figure 2.9) begins by mixing at least two constituents, the base material and the alkaline activator. The decision to choose the base material is influenced by several factors, such as its cost or availability, as well as the scope of application of the resulting geopolymer.

In most studies, the alkaline activator used is a solution that combines the dissolving ability of sodium hydroxide with the aggregating ability of sodium silicate. After mixing the components, the process is followed by a curing period at normal ambient conditions ($\approx 20°C$) or at slightly elevated temperatures ($<100°C$). During the curing stage, the chemical reaction of geopolymerization takes place. In the first step, the solid component is dissolved due to the presence of the activation solution. After removing a small amount of water, reorientation begins, where groups of atoms take their position in the structure and, at the same time, outline the solid structure of the geopolymer. In addition, the water is almost completely removed, and the material is transformed into its final form.

Since the geopolymerization reaction is time-governed, after the end of the curing step, it continues at the micro/nanoscopic level due to the reaction between the unreacted particles and the gel pore solution. This phenomenon gives geopolymers the ability to repair (self-heal) some of the cracks formed by dehydration (Abd Elhakam et al. 2012; Tang and Tang 2022; Liu et al. 2017; Kan et al. 2019).

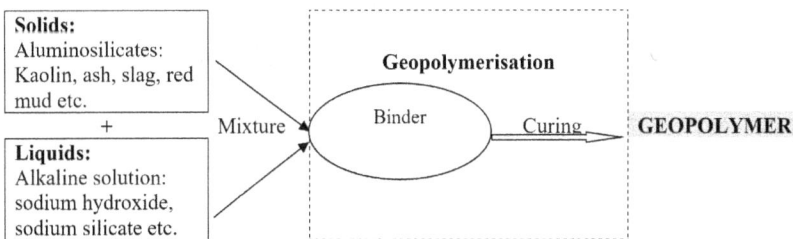

FIGURE 2.9 Schematic representation of the process of obtaining a polymer (Vizureanu and Burduhos Nergis 2020).

2.6 CLASSIFICATION

Geopolymers have a wide range of potential applications, including construction, waste management, and environmental remediation. Some of the main application areas for geopolymers include:

- Buildings: Can be used as a low-carbon alternative to traditional cement and concrete. They can be used to make lightweight concrete, insulation materials, and binders in building materials.
- Waste management: They can be manufactured from industrial waste such as fly ash and slag, making them a sustainable and cost-effective solution for managing these materials.
- Environmental remediation: They can be used to encapsulate and stabilize hazardous wastes such as heavy metals, radioactive materials, and organic pollutants.
- Fire retardant: They are known for their fire retardant properties and can be used to make fire retardant coatings, insulation, and fire retardant boards.
- Ceramics and refractories: Can be used to make ceramics, refractories, and other high-temperature applications.
- Biomedical: Some geopolymers are biocompatible and can be used to make bone cements, dental fillings, and other biomedical applications.

It's worth noting that these applications are in research and development and not yet widely commercialized.

Geopolymers comprise the following molecular units (or chemical groups):

- –Si–O–Si–O-siloxo, poly(siloxo)
- –Si–O–Al–O-sialate, poly(sialate)
- –Si–O–Al–O–Si–O-sialate-siloxo, poli(sialat-siloxo)
- –Si–O–Al–O–Si–O–Si–O-sialate-disiloxo, poly(sialate-disiloxo)
- –P–O–P–O– phosphate, poly(phosphate)
- –P–O–Si–O–P–O-phospho-siloxo, poly(phospho-siloxo)
- –P–O–Si–O–Al–O–P–O-phospho-sialate, poly(phospho–sialate)
- –(R)–Si–O–Si–O–(R) organo-siloxo, poly-silicone
- –A–O–P–O-alumino-phospho, poly(alumino-phospho)
- –Fe–O–Si–O–Al–O–Si–O-fero-sialate, poly(fero-sialate)

Geopolymers are currently being developed and applied to 10 main classes of materials:

- Geopolymer based on water glass, poly(siloxonate), soluble silicate, Si:Al = 1:0
- Geopolymer based on kaolinite/hydrosodalite, poly(sialate) Si:Al = 1:1
- Geopolymer based on metakaolin MK-750, poly(sialate-siloxo) Si:Al = 2:1
- Geopolymer based on calcium, (Ca, K, Na)-sialate, Si:Al = 1, 2, 3
- Rock-based geopolymer, poly(sialate-multisiloxo) 1< Si:Al<5
- Silica-based geopolymer, sialate link, and siloxo link in poly(siloxonate) Si:Al>5

- Geopolymer based on fly ash
- Geopolymer based on ferro-sialate
- Phosphate-based geopolymer, $AlPO_4$-based geopolymer
- Organic-mineral geopolymer

In Table 2.3, the classification according to the Si:Al ratio is presented. In Table 2.4, the classification of geopolymers according to type, activator, precursor, and respective applicability is presented.

TABLE 2.3
Classification of Geopolymers According to Ratio of Si:Al

Si:Al Ratio	Application
1	• Bricks • Ceramics • Materials for fire protection
2	• Cements and concretes with a small CO_2 footprint • Encapsulation of toxic and radioactive waste
3	• Fiberglass composites for fire protection • Equipment for foundries • Fire-resistant composites (200–1,000°C) • Tools for aeronautics—titanium processing
> 3	• Sealant for industry (200–600°C) • Tools for aeronautics—superplastic aluminum alloys
20–35	• Fire and temperature resistant composites

TABLE 2.4
General Classification of Geopolymers

Type	Polymeric Chain	Activation Solution	Precursors	Application
Sialates	Polysialat		Kaolin	Bricks
			Kaolinite	Ceramics
	Polysialat-siloxo	Natural	Metakaolin	Cement/Concrete
		Alkaline	Zeolites	Immobilization of hazardous waste
	Polysialat-disiloxo	Synthetic	Silica fume	Casting equipment
			Alumina	Aeronautic equipment
Phosphates	Phosphates	Natural	Zeolites	Membrane supports
			Kaolin	Insulating materials
	Phosphates-siloxo	Acid	Metakaolin	Refractory coatings
	Phosphates-sialate-siloxo	Synthetic	Silica fume	Monolithic roofs
			Alumina	Foams

(Continued)

TABLE 2.4 (*Continiued*)
General Classification of Geopolymers

Type	Polymeric Chain	Activation Solution	Precursors		Application
Iron sialates	Iron poly-sialates				Immobilization of hazardous waste
			Red mud		Bricks
					Binders and cements
		Alkaline			Sculptures
					Adsorption of heavy metals
			Slag		Plaster/concrete
					Building materials and flooring
Organic/ inorganic	Poly-organo-siloxo	Alkaline and acid	Organic polymeric network	Petroleum	
				Polivinilic Alcohol	High compression resistant concrete
				Epoxy	Immobilization of hazardous waste
	Kerogen			Lignine	Gas or oil source
				Humic acid	Concrete with high flexural strength
				Metakaolin	
				Kaolin	

From these tables, it can be observed that geopolymers can be classified in several groups, firstly by the ratio of Si/Al and also according to the type and polymeric chain, in natural and synthetic ones, based on sialates, phosphates, iron, and organic/ inorganic ones. Their applications vary from general construction to special applications such as the immobilization of hazardous waste or insulating materials.

2.7 CONCLUSIONS

Geopolymers are a type of inorganic polymer that is formed through the reaction of an aluminosilicate source material, such as fly ash or metakaolin, with an alkaline activator solution, typically a mixture of sodium hydroxide and sodium silicate. The resulting material is a high-strength, durable material that can be used in a variety of applications, from construction to industrial manufacturing.

One of the key advantages of geopolymers is their environmental sustainability. They are typically made from waste materials such as fly ash, which would otherwise be disposed of in landfills. In addition, the production of geopolymers generates significantly less CO_2 emissions compared to traditional cement manufacturing processes.

Geopolymers have a wide range of applications, including as a replacement for traditional cement in construction, as a high-temperature insulation material, and as a binder in the production of refractory materials. They have also been used in the development of new materials such as geopolymer-based composites.

Overall, geopolymers have shown great promise as a sustainable alternative to traditional cement and other materials, and their use is expected to grow in the coming years as more research is conducted into their properties and applications.

REFERENCES

Abbas, A.G.N., F.N.A.A. Aziz, K. Abdan, N.A.M. Nasir, and G.F. Huseien. 2022. A State-of-the-Art Review on Fibre-Reinforced Geopolymer Composites. *Construction and Building Materials* 330: 127187.

Abd Elhakam, A., A.E. Mohamed, and E. Awad. 2012. Influence of Self-Healing, Mixing Method and Adding Silica Fume on Mechanical Properties of Recycled Aggregates Concrete. *Construction and Building Materials* 35: 421–427.

Abdila, S.R., M.M.A.B. Abdullah, R. Ahmad, D.D.B. Nergis, S.Z.A. Rahim, M.F. Omar, A.V. Sandu, P. Vizureanu, and S. Syafwandi. 2022. Potential of Soil Stabilization Using Ground Granulated Blast Furnace Slag (GGBFS) and Fly Ash via Geopolymerization Method: A Review. *Materials* 5, no. 1: 375. https://www.mdpi.com/1996-1944/15/1/375/htm.

Abdollahnejad, Z., S. Miraldo, F. Pacheco-Torgal, and J.B. Aguiar. 2017. Cost-Efficient One-Part Alkali-Activated Mortars with Low Global Warming Potential for Floor Heating Systems Applications. *European Journal of Environmental and Civil Engineering* 21, no. 4: 412–429.

Ahmad, J., D.D. Burduhos-Nergis, M.M. Arbili, S.M. Alogla, A. Majdi, and A.F. Deifalla. 2022. A Review on Failure Modes and Cracking Behaviors of Polypropylene Fibers Reinforced Concrete. *Buildings* 12, no. 11: 1951. https://www.mdpi.com/2075-5309/12/11/1951/htm.

Andrew, R.M. 2018. Global CO_2 Emissions from Cement Production, 1928-2017. *Earth System Science Data* 10, no. 4: 2213–2239.

Bell, J.L., P.E. Driemeyer, and W.M. Kriven. 2009. Formation of Ceramics from Metakaolin-Based Geopolymers. Part II: K-Based Geopolymer. *Journal of the American Ceramic Society* 92, no. 3: 607–615.

Belviso, C. 2018. State-of-the-Art Applications of Fly Ash from Coal and Biomass: A Focus on Zeolite Synthesis Processes and Issues. *Progress in Energy and Combustion Science* 65: 109–135.

Burduhos Nergis, D.D., M.M.A.B. Abdullah, P. Vizureanu, and M.F. Mohd Tahir. 2018. Geopolymers and Their Uses: Review. *IOP Conference Series: Materials Science and Engineering* 374, no. 1: 012019.

Çelik, A.İ., Y.O. Özkılıç, Ö. Zeybek, N. Özdöner, and B.A. Tayeh. 2022. Performance Assessment of Fiber-Reinforced Concrete Produced with Waste Lathe Fibers. *Sustainability (Switzerland)* 14, no. 19. doi:10.3389/fmats.2022.1057128

Cong, P. and Y. Cheng. 2021. Advances in Geopolymer Materials: A Comprehensive Review. *Journal of Traffic and Transportation Engineering* (English Edition) 8, no. 3: 283–314.

Dai, S., H. Wang, S. An, and L. Yuan. 2022. Mechanical Properties and Microstructural Characterization of Metakaolin Geopolymers Based on Orthogonal Tests. *Materials* 15, no. 8: 2957. https://www.mdpi.com/1996-1944/15/8/2957/htm.

Davidovits, J. 1991. Geopolymers: Inorganic Polymeric New Materials. *Journal of Thermal Analysis* 37, no. 8: 1633–1656.

Davidovits, J. 2015. *False-CO$_{2-Values}$* (Technical Paper #24, Scientific Papers). Geopolymer Institute Library, www.Geopolymer.Org. https://www.materialstoday.com/polymers-soft-materials/features/environmental-implications-of-geopolymers/,.

Deb, P.S., P. Nath, and P.K. Sarker. 2014. The Effects of Ground Granulated Blast-Furnace Slag Blending with Fly Ash and Activator Content on the Workability and Strength Properties of Geopolymer Concrete Cured at Ambient Temperature. *Materials and Design* 62: 32–39.

Gartner, E. and H. Hirao. 2015. A Review of Alternative Approaches to the Reduction of CO_2 Emissions Associated with the Manufacture of the Binder Phase in Concrete. *Cement and Concrete Research* 78: 126–142.

He, Z., X. Zhu, J. Wang, M. Mu, and Y. Wang. 2019. Comparison of CO_2 Emissions from OPC and Recycled Cement Production. *Construction and Building Materials* 211: 965–973.

Heah, C.Y., H. Kamarudin, A.M. Mustafa Al Bakri, M. Bnhussain, M. Luqman, I.K. Nizar, C.M. Ruzaidi, and Y.M. Liew. 2012. Study on Solids-to-Liquid and Alkaline Activator Ratios on Kaolin-Based Geopolymers. *Construction and Building Materials* 35: 912–922.

Isa, A., N. Nosbi, M.C. Ismail, H.M. Akil, W.F.F. Wan Ali, and M.F. Omar. 2022. A Review on Recycling of Carbon Fibres: Methods to Reinforce and Expected Fibre Composite Degradations. *Materials* 15, no. 14: 4991. https://www.mdpi.com/1996-1944/15/14/4991/htm.

Provis, J.L. and J.S.J. van Deventer. 2007. *Geopolymers Structures, Processing, Properties and Industrial Applications* (Vol. 1). Woodhead Publishing Series, London.

Jat, D., R. Motiani, S. Dalal, and I. Thakar. 2023. Mechanical Properties of Geopolymer Concrete Reinforced with Various Fibers: A Review. In: *Proceedings of the 2nd International Symposium on Disaster Resilience and Sustainable Development* (pp. 139–156). Springer, New Nork, NY. doi:10.1007/978-981-19-6297-4_11.

Jiang, C., A. Wang, X. Bao, T. Ni, and J. Ling. 2020. A Review on Geopolymer in Potential Coating Application: Materials, Preparation and Basic Properties. *Journal of Building Engineering* 32: 101734.

Joseph, D. 2008. *Geopolymer Chemistry and Applications* (Vol. 1, 5th edn). Geopolymer Institute, Saint-Quentin. https://www.researchgate.net/publication/265076752.

Kan, L., J. Wei, B. Duan, and M. Wu. 2019. Self-Healing of Engineered Geopolymer Composites Prepared by Fly Ash and Metakaolin. *Cement and Concrete Research* 125: 105895.

Kanagaraj, B., E. Lubloy, N. Anand, V. Hlavicka, and T. Kiran. 2023. Investigation of Physical, Chemical, Mechanical, and Microstructural Properties of Cement-Less Concrete: State-of-the-Art Review. *Construction and Building Materials* 365: 130020.

Kenne Diffo, B.B., A. Elimbi, M. Cyr, J.D. Manga, and H.T. Kouamo. 2015. Effect of the Rate of Calcination of Kaolin on the Properties of Metakaolin-Based Geopolymers. *Journal of Asian Ceramic Societies* 3, no. 1: 130–138.

Khale, D. and R. Chaudhary. 2007. Mechanism of Geopolymerization and Factors Influencing Its Development: A Review. *Journal of Materials Science* 42, no. 3: 729–746.

Khalifeh, M., A. Saasen, H. Hodne, R. Godøy, and T. Vrålstad. 2018. Geopolymers as an Alternative for Oil Well Cementing Applications: A Review of Advantages and Concerns. *Journal of Energy Resources Technology, Transactions of the ASME* 140: 9.

Koleżyński, A., M. Król, and M. Żychowicz. 2018. The Structure of Geopolymers: Theoretical Studies. *Journal of Molecular Structure* 1163: 465–471.

Komnitsas, K. and D. Zaharaki. 2007. Geopolymerisation: A Review and Prospects for the Minerals Industry. *Minerals Engineering* 20, no. 14: 1261–1277.

Li, X., Z. Yu, B. Ma, and B. Wu. 2010a. Effect of MSWI Fly Ash and Incineration Residues on Cement Performances. *Journal Wuhan University of Technology, Materials Science Edition* 25, no. 2: 312–315.

Li, C., H. Sun, and L. Li. 2010b. A Review: The Comparison between Alkali-Activated Slag (Si + Ca) and Metakaolin (Si + Al) Cements. *Cement and Concrete Research* 40, no. 9: 1341–1349.

Liew, Y.-M., C.-Y. Heah, A.B.M. Mustafa, and H. Kamarudin. 2016. Structure and Properties of Clay-Based Geopolymer Cements: A Review. *Progress in Materials Science* 83: 595–629. doi:10.1016%2Fj.pmatsci.2016.08.002.

Liu, X., M.J. Ramos, S.D. Nair, H. Lee, D.N. Espinoza, and E. van Oort. 2017. True Self-Healing Geopolymer Cements for Improved Zonal Isolation and Well Abandonment. In: *SPE/IADC Drilling Conference, Proceedings*, March 2017, The Hague, The Netherlands (pp. 130–141).

Maichin, P., P. Jitsangiam, T. Nongnuang, K. Boonserm, K. Nusit, S. Pra-Ai, T. Binaree, and C. Aryupong. 2021. Stabilized High Clay Content Lateritic Soil Using Cement-FGD Gypsum Mixtures for Road Subbase Applications. *Materials* 14, no. 8: 1858. https://www.mdpi.com/1996-1944/14/8/1858/htm.

Mehmood, A., M. Irfan-ul-Hassan, and N. Yaseen. 2022. Role of Industrial By-Products and Metakaolin in the Development of Sustainable Geopolymer Blends: Upscaling from Laboratory-Scale to Pilot-Scale. *Journal of Building Engineering* 62: 105279.

Nie, Q., Y. Li, G. Wang, and B. Bai. 2020. Physicochemical and Microstructural Properties of Red Muds under Acidic and Alkaline Conditions. *Applied Sciences* 10, no. 9: 2993. https://www.mdpi.com/2076-3417/10/9/2993/htm.

Nikolov, A., H. Nugteren, and I. Rostovsky. 2020. Optimization of Geopolymers Based on Natural Zeolite Clinoptilolite by Calcination and Use of Aluminate Activators. *Construction and Building Materials* 243: 118257. doi:10.1016/j.conbuildmat.2020.118257.

Nowak, A., M. Lubas, J.J. Jasinski, M. Szumera, R. Caban, J. Iwaszko, and K. Koza. 2022. Effect of Dolomite Addition on the Structure and Properties of Multicomponent Amphibolite Glasses. *Materials* 15, no. 14: 4870. https://www.mdpi.com/1996-1944/15/14/4870/htm.

Papa, E., V. Medri, S. Amari, J. Manaud, P. Benito, A. Vaccari, and E. Landi. 2018. Zeolite-Geopolymer Composite Materials: Production and Characterization. *Journal of Cleaner Production* 171: 76–84. doi:10.1016%2Fj.jclepro.2017.09.270.

Papa, E., V. Medri, C. Paillard, B. Contri, A.N. Murri, A. Vaccari, and E. Landi. 2019. Geopolymer-Hydrotalcite Composites for CO_2 Capture. *Journal of Cleaner Production* 237: 117738.

Provis, J.L. 2009. Activating Solution Chemistry for Geopolymers. *Geopolymers: Structures, Processing, Properties and Industrial Applications* 8: 50–71.

Qaidi, S.M.A., B.A. Tayeh, A.M. Zeyad, A.R.G. de Azevedo, H.U. Ahmed, and W. Emad. 2022. Recycling of Mine Tailings for the Geopolymers Production: A Systematic Review. *Case Studies in Construction Materials* 16: e00933

Qureshi, M.N. and S. Chish. 2013. Workability and Setting Time of Alkali Activated Blast Furnace Slag Paste. *Advances in Civil Engineering Materials* 2, no. 1: 62–77. https://www.iosrjournals.org/.

Rahier, H., J. Wastiels, M. Biesemans, R. Willlem, G. van Assche, and B. van Mele. 2007. Reaction Mechanism, Kinetics and High Temperature Transformations of Geopolymers. *Journal of Materials Science* 42, no. 9: 2982–2996.

Ranjbar, N. and M. Zhang. 2020. Fiber-Reinforced Geopolymer Composites: A Review. *Cement and Concrete Composites* 107: 103498.

Rowles, M. and B. O'Connor. 2003. Chemical Optimisation of the Compressive Strength of Aluminosilicate Geopolymers Synthesised by Sodium Silicate Activation of Metakaolinite. *Journal of Materials Chemistry* 13, no. 5: 1161–1165.

Rożek, P., M. Król, and W. Mozgawa. 2019. Geopolymer-Zeolite Composites: A Review. *Journal of Cleaner Production* 230: 557–579.

Sathonsaowaphak, A., P. Chindaprasirt, and K. Pimraksa. 2009. Workability and Strength of Lignite Bottom Ash Geopolymer Mortar. *Journal of Hazardous Materials* 168, no. 1: 44–50.

Skariah Thomas, B., J. Yang, A. Bahurudeen, S.N. Chinnu, J.A. Abdalla, R.A. Hawileh, and H.M. Hamada. 2022. Geopolymer Concrete Incorporating Recycled Aggregates: A Comprehensive Review. *Cleaner Materials* 3: 100056.

Sotelo-Piña, C., E.N. Aguilera-González, and A. Martínez-Luévanos. 2018. Geopolymers: Past, Present, and Future of Low Carbon Footprint Eco-Materials. In: *Handbook of Ecomaterials* (pp. 1–21). Springer, New York, NY. https://link.springer.com/referenceworkentry/10.1007/978-3-319-48281-1_54-1.

Tang, D. and H. Tang. 2022. Self-Healing Diamond/Geopolymer Composites Fabricated by Extrusion-Based Additive Manufacturing. *Additive Manufacturing* 56: 102898.

Taylor, P., A.I. Journal, F. Rao, and Q. Liu. 2015. Geopolymerization and Its Potential Application in Mine Tailings Consolidation : A Review Geopolymerization and Its Potential Application in Mine Tailings Consolidation : A Review. *Mineral Processing and Extractive Metallurgy Review* 36: 399–409.

Tho-In, T., V. Sata, K. Boonserm, and P. Chindaprasirt. 2016. Compressive Strength and Microstructure Analysis of Geopolymer Paste Using Waste Glass Powder and Fly Ash. *Journal of Cleaner Production* 172: 2892–2898.

van Deventer, J.S.J., J.L. Provis, and P. Duxson. 2012. Technical and Commercial Progress in the Adoption of Geopolymer Cement. *Minerals Engineering* 29: 89–104.

Vizureanu, P. and D.D. Burduhos Nergis. 2020. *Green Materials Obtained by Geopolymerization for a Sustainable Future* (Vol. 90, p. 105). Materials Research Foundations, Millersville, PA. https://books.google.ro/books?hl=ro&lr=&id=GccLEAAAQBAJ&oi=fnd&pg=PP4&ots=QEyq5WN5Z_&sig=SCSckNzhyW-q1BMJKbNt1q5dy7E&redir_esc=y#v=onepage&q&f=false.

Wang, H., H. Li, and F. Yan. 2005. Synthesis and Mechanical Properties of Metakaolinite-Based Geopolymer. *Colloids and Surfaces A: Physicochemical and Engineering Aspects* 268, no. 1–3: 1–6.

Wang, Y.S., Y. Alrefaei, and J.G. Dai. 2021. Roles of Hybrid Activators in Improving the Early-Age Properties of One-Part Geopolymer Pastes. *Construction and Building Materials* 306: 124880

Xu, H. and J.S.J. van Deventer. 2000. The Geopolymerisation of Alumino-Silicate Minerals. *International Journal of Mineral Processing* 59, no. 3: 247–266.

Xue, C., V. Sirivivatnanon, A. Nezhad, and Q. Zhao. 2023. Comparisons of Alkali-Activated Binder Concrete (ABC) with OPC Concrete: A Review. *Cement and Concrete Composites* 135: 104851.

Yan, S., P. He, D. Jia, Z. Yang, X. Duan, S. Wang, and Y. Zhou. 2016. Effect of Fiber Content on the Microstructure and Mechanical Properties of Carbon Fiber Felt Reinforced Geopolymer Composites. *Ceramics International* 42, no. 6: 7837–7843.

Yao, X., Z. Zhang, H. Zhu, and Y. Chen. 2009. Geopolymerization Process of Alkali-Metakaolinite Characterized by Isothermal Calorimetry. *Thermochimica Acta* 493, no. 1–2: 49–54.

Yip, C.K., G.C. Lukey, and J.S.J. van Deventer. 2005. The Coexistence of Geopolymeric Gel and Calcium Silicate Hydrate at the Early Stage of Alkaline Activation. *Cement and Concrete Research* 35, no. 9: 1688–1697.

Zhang, H.Y., X. Hao, and W. Fan. 2016. Experimental Study on High Temperature Properties of Carbon Fiber Sheets Strengthened Concrete Cylinders Using Geopolymer as Adhesive. *Procedia Engineering* 135: 47–55.

Zhang, J., Y. Ge, Z. Li, and Y. Wang. 2020a. Facile Fabrication of a Low-Cost and Environmentally Friendly Inorganic-Organic Composite Membrane for Aquatic Dye Removal. *Journal of Environmental Management* 256: 109969. doi:10.1016/j.jenvman.2019.109969.

Zhang, P., K. Wang, Q. Li, J. Wang, and Y. Ling. 2020b. Fabrication and Engineering Properties of Concretes Based on Geopolymers/Alkali-Activated Binders: A Review. *Journal of Cleaner Production* 258. doi:10.1016/j.jclepro.2020.120896

Ziada, M., S. Erdem, Y. Tammam, S. Kara, and R.A.G. Lezcano. 2021. The Effect of Basalt Fiber on Mechanical, Microstructural, and High-Temperature Properties of Fly Ash-Based and Basalt Powder Waste-Filled Sustainable Geopolymer Mortar. *Sustainability* 13, no. 22: 12610. https://www.mdpi.com/2071-1050/13/22/12610/htm.

Zribi, M. and S. Baklouti. 2021. Phosphate-Based Geopolymers: A Critical Review. *Polymer Bulletin.* doi:10.1007/s00289-021-03829-0

3 Geopolymer-Reinforced Steel Fibers

Meor Ahmad Faris Meor Ahmad Tajudin,
Nurul Aida Mohd Mortar, and
Muhammad Faheem Mohd Tahir
Universiti Malaysia Perlis

Dumitru Doru Burduhos-Nergis
Gheorghe Asachi Technical University

Poppy Puspitasari
Universitas Brawijaya

3.1 INTRODUCTION

The development of non-cementitious concrete, known as geopolymer concrete, has increased worldwide demand for materials in the construction industry. Geopolymer concrete consists of any raw materials containing silica (SiO_2) and alumina (Al_2O_3) in high compositions, which are mixed with alkali-activated binder, crushed stone, and sand. The main ingredients of fly ash, which are rich in alumina and silica, can boost the rate of geopolymerization. Geopolymerization is the result of the polymerization process of aluminosilicate materials, silica (Si) and alumina (Al), in a high alkaline solution such as sodium hydroxide (Temuujin et al., 2013). The geopolymerization consists of three stages: dissolution of the aluminosilicate, gel formation, and polycondensation (Okoye et al., 2017). Thus, the formation of the geopolymer binder known as aluminosilicate gel provides a superior effect compared to ordinary Portland cement (OPC).

Geopolymer concrete yielded good performance in terms of mechanical and physical properties when compared to OPC concrete. The geopolymer concrete failure mode and behavior are similar to conventional concrete. Concrete was considered a brittle material with good compression but less resistance to tension (Ranjbar et al., 2016). The improvement of concrete weakness can be overcome by adding reinforcement bars, which increase the strain capacity and bending strength. Aswani and Karthi (2017) stated that reinforced concrete can normally increase its tensile strength by up to 10%. As part of the revolution toward solving the tensile strength problem in concrete, fibers were proposed to improve its durability. There are many further kinds of research being done to develop the design of reinforcement structures in concrete, which can establish an effective and economic standard for reinforced geopolymer concrete.

DOI: 10.1201/9781003390190-3

3.2   REINFORCEMENT IN GEOPOLYMER CONCRETE

The addition of reinforcement in geopolymer concrete is steel and fiber, which have different functions, arrangements, and capabilities to resist internal or external loading.

There are two major types of reinforcement used to reinforce concrete: steel bar and fiber. The further result of reinforcing the steel bar in concrete composites can improve the shear toughness and flexural strength of the composites (Faris et al., 2016). In the lab work, Sarker and McBeath (2015) chose deformed steel bars with two different diameters, 10 and 12 mm, which yield strengths of 560 and 520 MPa, respectively. Due to that, the increased tensile strength of geopolymer concrete allows for the development of a larger bending strength that be produced prior to the formation of cracking (Nematollahi et al., 2017). Figure 3.1 shows the cross-section view of a geopolymer concrete beam with steel rebar grade B500BT with a nominal yield of 500 MPa. The arrangement of the longitudinal reinforcement can reduce the damage degree to concrete structure, and an optimal shear reinforcement arrangement is also able to minimize the ballistic resistance (Zhao & Chen, 2013).

Furthermore, the incorporation of fiber reinforcement materials such as steel, glass, nylon, carbon, polypropylene, and polyvinyl alcohol can increase the geopolymer concrete's capability to resist impact and shear stress effects. The orientation of fiber was randomly distributed in a geopolymer concrete, as shown in Figure 3.2.

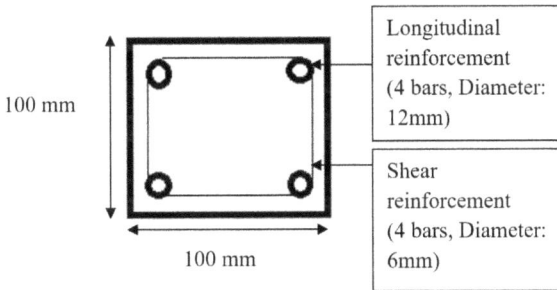

FIGURE 3.1   Arrangement of steel bar reinforcement.

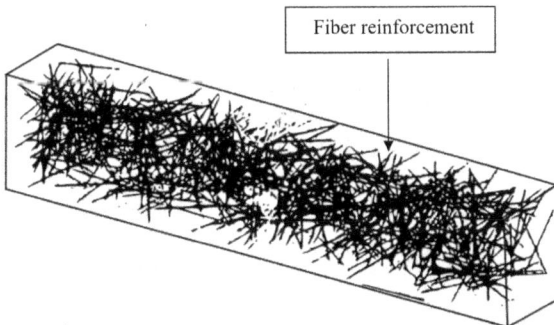

FIGURE 3.2   Fiber orientation in geopolymer concrete.

According to Sanjayan et al. (2015), when fibers are properly distributed among the matrix, it effectively intercepts the micro-cracks and evenly slows down the interaction of micro-cracks, thereby providing less stress transfer growth. After the pre-crack appears, fiber-reinforced concrete improves the toughness of the structure and maintains the ability to reveal high tensile ductility at the weak stress concentration point.

3.3 FIBERS REINFORCEMENT IN GEOPOLYMER CONCRETE

The inclusion of different fiber types and different fiber ratios examined the cracking behavior of geopolymer composites. It is an important parameter to characterize the performance of different fiber combinations used to reinforce the brittle matrix to resist the load capacity.

3.3.1 INCLUSION OF DIFFERENT FIBER MATERIALS

The general types of fiber material used during the mix proportion to enhance the strength were steel fiber, polymeric fiber, and natural fiber. Nowadays, studies on steel fiber are widely investigated to build hybrid high-performance concrete. The basic properties of steel fiber indicate that the addition of steel fiber exhibits high strength and stiffness, and the load can increase more quickly than with polymer fibers. There are different shapes of fiber, for instance, straight, hook, and crimped shapes. The hooked-end steel fiber is the better fiber among the steel fibers because of the hook shape that is capable of bonding with the geopolymer matrix (Soetens et al., 2014).

Regarding other fiber reinforcement, polymer fiber can be used alternatively. There is polyvinyl alcohol (PVA), polyvinyl chloride (PVC), and polypropylene (PP). Alomayri et al. (2014) studied PVA fibers, which exhibit better post-crack propagation and strain hardening behavior in comparison to PVC, yet they rarely have a high enough stiffness to maximize the pre-crack control toward the ultimate load. PVC-reinforced concrete results in a more homogenous and uniform distribution compared to other polymer fibers. However, PP fiber has a lower Young's modulus but higher elongation at break than PVA and PVC fiber and brings more cost efficiency.

Concerning alternative natural-based reinforcement, the behavior of fibers such as wool, sweet sorghum, cellulose, coconut, and wood flour is reproducible, has high specific strength, low density, and is cheap to obtain. The addition of natural fiber, such as wool fiber agglomeration, was observed at a lower ratio, which degraded the interfacial adhesion between the fiber and the matrix (Korniejenko et al., 2016). Coconut is one of the most common foods and industrial plants. The coir fibers can be extracted from the coconut husk mechanically or manually. The raffia palm tree is very useful and known as a multifunctional plant family. Its nuts are a source of food and cosmetic oil, whereas the petioles and raw leaves are used as construction materials, and the raw fibers are extracted from the upper surface of the leaflets.

3.3.2 Inclusion of Different Fiber Content

The amount of fiber content in geopolymer concrete was determined to reach the optimum performance of fracture toughness in fiber-reinforced geopolymer composites. The amount of fibers that are usually added to the concrete mix is calculated in percentages from the total volume of the concrete. Al-Majidi et al. (2017) investigated tensile and flexural strength by adding steel fibers in the volume range of 0.5%, 1.0%, 1.5%, and 2.0%, respectively. The result indicates that the compressive strength of fiber concrete reached a maximum of 1.5% volume fraction, representing a 15.3% improvement over the concrete without reinforcement. The flexural strength and split tensile improved to 98.3% and 126.6% at a 2.0% volume fraction, respectively. If the fiber volume fraction is sufficiently high, this results in a geopolymer concrete increase in the tensile strength and capability of absorbing energy due to applied loads (Guo & Pan, 2018).

In terms of fracture properties, Alomayri et al. (2013) investigated the mechanical and fracture behavior of geopolymer mixed with cotton fiber at 0.3%–1.0% by volume fraction. They found the optimum cotton fiber content is around 0.5%, which provides the highest flexural strength and fracture toughness of about 11.7 and 1.12 MPa, respectively. The geopolymer-reinforced specimens containing steel fibers have higher strength than the geopolymer containing polymer contamination. Increasing the amount of fiber has a positive effect on the development of strength. Steel fiber reinforcement in a 1% volume fraction achieved significant growth in compressive strength and three times the growth in flexural strength as compared to unreinforced geopolymer concrete.

3.4 PHYSICAL PROPERTIES OF STEEL FIBER-REINFORCED GEOPOLYMER CONCRETE

Physical of geopolymer concrete related to workability, density, water absorption, and porosity were analyzed, and details will be discussed in this section.

3.4.1 Effects of Steel Fiber Addition on Slump Test

The fresh geopolymer concrete is normally measured by the conventional slump test due to its being easy to handle and very commonly used in practice. Fresh geopolymer concrete normally appears dark and shiny. Besides, a slump is very useful to analyze the consistency of concrete, in which variations in the uniformity of mixtures can be detected.

There are many factors influencing the workability of concrete, such as water content, size of aggregates, surface texture of aggregates, admixtures, mix proportions, shape of aggregates, and grading of aggregates. Besides, the inclusion of fibers also has an impact on the workability of concrete. Fibers in geopolymer concrete act as rigid additions with a certain surface area and are different from coarse aggregate in geometry. Fibers' dimensions, such as length, diameter, and shape that have a higher surface area give a higher probability of increasing the workability of fresh concrete. The addition of fibers is normally decided based on slump production to ensure excellent results of the steel fiber-reinforced geopolymer concrete (SFRGC) product at a specific application. However, Narayanan and Kareem-Palanjian (1982) state that the optimum fiber content is decided when fiber balling takes place.

The influence of fibers on the workability of geopolymer concrete is mainly due to the shape of fibers, which are more elongated compared to aggregates and create interlocking. The higher surface area of fibers results in a higher water demand. Besides, the surface properties of fibers also play an important role, as the surface of fibers is different from that of cement and aggregate. The surface of fibers might be hydrophilic or hydrophobic. Excellent bond strength in the fiber-geopolymer matrix is produced from fibers that promote a hydrophilic surface. Ranjbar et al. (2016) have proven that the surface texture of steel fibers is rougher than polypropylene, thereby producing a better interfacial bond strength between fiber-geopolymer matrix. This result was proven to be consistent with flexural strength results. The shape of fibers, such as hooked, straight, crimped, and mechanically deformed, also affects the workability of geopolymer concrete. This different shape was innovated to improve the anchorage between the fibers and the geopolymer matrix. Hence, friction between fiber-geopolymer matrices will get higher. For example, steel fibers with hooked ends have higher workability than straight steel fibers. Besides, the additional higher amount of fibers reduces the workability of geopolymer concrete.

3.4.2 Effects of Steel Fiber Addition on Density

Based on a previous study, the density of standard geopolymer concrete is about 2,400 kg/m³. This value is similar to the density of OPC concrete. The density of geopolymer concrete increases with the increase in steel fiber content. This trend was stated in a previous study where Shafigh et al. (2011) proved the inclusion of steel fibers will slightly increase the density of concrete. This is due to the high specific value of steel fibers, which is 7,850 kg/m³.

A relationship with a strong correlation for concrete samples has been proposed. In this study, Shafigh et al. (2011) proposed an equation that represents the density of a concrete sample as below:

$$D_d = 88_f^2 + 14V_f + 1940.6 \qquad (3.1)$$

where D_d is demoulded density (kg/m³), D_a is air-dry density (kg/m³), and V_f is fiber volume.

The impact of steel fibers addition to the density of concrete will be higher if the density of concrete is low, such as in light weight concrete. The addition of steel fibers to lightweight concrete has been done by Hassanpour et al. (2014), where the impact on the density of the concrete was significant. However, additions of fibers to lightweight concrete are critical due to their increased density, which is not suitable for lightweight applications that aim to reduce weight as much as possible. Limited steel fiber addition was proposed to reduce the impact on the density of concrete, where the maximum addition suggested is 1% by volume.

3.4.3 Effects of Steel Fiber Addition on Water Absorption

Water absorption of standard geopolymer concrete normally confronts ASTM C140, in which the average water absorption of geopolymer concrete samples shall not be greater than 5%, with no individual unit greater than 7%. This shows that the properties of geopolymer concrete have excellent behavior in terms of water absorption. Hence, geopolymer concrete is not highly permeable, and water absorption below 5% is classified as low permeable. This is supported by Rendell et al. (2002). Low permeability represents low porous concrete. This was agreed upon by Gunasekara et al. (2016).

In its most fundamental form, permeability is the degree to which water, air, or any other substance, such as an ion of chloride, penetrates inside the concrete. Small pores in geopolymer concrete allow any possible substances to be absorbed inside the concrete. Higher water absorption may result from the higher porousness of geopolymer concrete. This shows that the water absorption of geopolymer concrete has a strong relationship with porosity.

Meanwhile, water absorption is also correlated with the durability of geopolymer concrete, where lower water absorption helps to reduce the possibility of steel fibers inside geopolymer concrete corroding. This prevention of corrosion by low water absorption occurred due to limiting the penetration of the chloride into the geopolymer concrete. A thin layer has formed surrounding the steel fibers, which naturally helps to protect them from corrosion attack. This thin layer, known as a passive layer, is produced from a chemical reaction between steel fibers and a matrix of geopolymer when there is contact between them. This chloride can penetrate geopolymer concrete through small pores and accumulate around the steel fibers that are already coated by a natural passive layer. After a certain period, the chloride will break down the passive layer, which will result in the steel fibers corroding.

Rust is a product of corrosion that takes up a larger volume than the original steel fibers that result from the concrete cracks when reinforcing steel fibers undergo a corrosion process. After that, the corrosion process will become faster due to the chloride flowing easily into direct contact with the steel fibers after the passive layer has broken down. As a result, the performance of SFRGC in terms of compressive strength, flexural strength, flexural toughness, interfacial bonding strength, and durability will drop. In conclusion, it is important to produce low water absorption in geopolymer concrete to minimize the reduction of SFRGC performance.

3.5 MECHANICAL PROPERTIES OF STEEL FIBERS-REINFORCED GEOPOLYMER CONCRETE

Effects of steel fiber addition on the mechanical properties of geopolymer concrete, including compressive strength, flexural strength, and flexural toughness, will be discussed in this section. The relative strength property of geopolymer concrete varies with the inclusion of steel fibers.

3.5.1 Effects of Steel Fiber Addition on Compressive Strength

In general, steel fibers help to improve the flexural strength of brittle matrixes and normally have a small effect on the compressive strength. The increase in compressive strength with the addition of fibers was reported by many authors from previous studies, including Al-mashhadani et al. (2018), Reed et al. (2014), and Lou et al. (2009). However, there are some studies reported that claim fiber additions reduce compressive strength due to the poor workability of certain fibers. The reduction of compressive strength by the addition of fibers was reported in a few studies. Enhancement of compressive strength exhibited by geopolymer samples with the addition of fibers can normally increase by up to 50%.

In compressive loads, cracks normally initiate at the coarse aggregate and then propagate to the cement binder. At this moment, steel fibers are responsible as crack growth arrestors to stop the crack propagation, hence increasing the ultimate compressive strength of concrete. This is agreed upon by Gao et al. (1997). The high stiffness of steel fibers, which have a high aspect ratio, promotes better control over micro-crack propagation. The arrest of micro-cracks resulted in a delay in the initiation of macro-cracks and led to an increase in compressive strength due to the fact that concrete can sustain a higher load. This ability of steel fibers as reinforcement to stop the micro- and macro-cracks in the pre-peak region makes this reinforcing fiber a contributor to increasing the compressive strength.

The inclusion of different types of fibers and amounts by volume percentage into the unreinforced geopolymer matrix can be analyzed by measuring a marginal response variation in terms of compressive strength results. The increase and decrease of compressive strength developed by geopolymer concrete samples with additions of fibers (at the same mix design as unreinforced concrete) are calculated based on the strength index that has been proposed by Khan et al. (2018) as in Equation (3.2) below:

$$\text{Strength index} = \frac{(\sigma_f - \sigma_m)}{\sigma_m} \times 100\% \qquad (3.2)$$

where is σ_f the mean compressive strength of the composite samples with fibers and σ_m is the mean compressive strength of the plain geopolymer matrix.

The failure pattern of unreinforced geopolymer concrete is different from the failure pattern of reinforced geopolymer concrete. Figure 3.3 shows the crack pattern of both unreinforced and reinforced geopolymer concretes. This happened due to the inclusion of steel fibers, which helped to stop the crack propagations at one point and initiate a new crack at a different region of the weak point with the addition of compression load.

3.5.2 Effects of Steel Fibers Addition on Flexural Strength

Generally, flexural strength is tested to evaluate the mechanical properties of concrete. This flexural strength study is conducted to study the direct tensile strength

FIGURE 3.3 Crack pattern of (a) unreinforced and (b) reinforced geopolymer concrete.

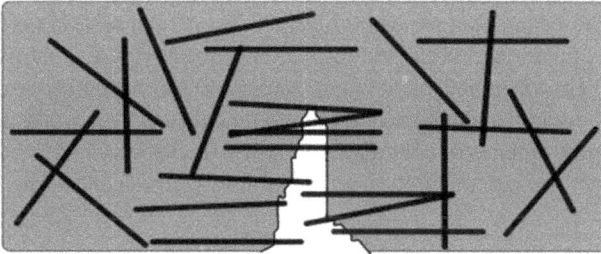

FIGURE 3.4 Schematic of bridging effect in geopolymer concrete.

and resistance to bending of concrete under an applied load. The most familiar test to evaluate flexural strength is known as four-point bending or third-point bending.

The inclusion of steel fibers in geopolymer concrete is very important for post-cracking improvement. The inclusion of fibers helps to reduce the brittleness of standard geopolymer concrete. The increase in flexural strength with the addition of fibers in geopolymer concrete was reported elsewhere. The improvement of flexural strength with the addition of fibers can increase by up to 200%. The increase in flexural strength of concrete samples was very influenced by fiber addition, where steel fibers functioned to create bridges between the crack spots, as seen in Figure 3.4. At this stage, the inclusion of fibers helps to change the concrete's behavior from brittle to ductile.

Unreinforced geopolymer concrete normally fails immediately after initial cracking. Meanwhile, reinforced geopolymer concrete will undergo post-cracking load-carrying capacity. This theory was agreed upon by Yoo et al. (2015), who claimed fiber inclusion in OPC concrete helps to bridge the crack spot if the total amount of fibers is sufficient. Steel fibers are proven to have the highest load-carrying capacity. This is proven by a previous study where Khan et al. (2018) concluded that steel fibers produce the highest load-carrying capacity compared to other types of fibers at the same volume fraction.

Different types of fiber ends normally have different effects on the flexural strength of geopolymer concrete. For example, the flexural strength of hooked steel fibers indicates a higher flexural strength compared to straight steel fibers. This is due to the hooked end of steel fibers having an anchorage that contributes to increasing the bond strength between the steel fibers and the geopolymer matrix. Extra load is needed to deform the hooked end of the steel fibers during the bridging effects where fibers are pulled out of the matrix. The hooked end then becomes a straight end when it is pulled from the matrix.

3.6 CONCLUSIONS

In conclusion, the addition of fibers is proven to improve the brittleness of geopolymer concrete without sacrificing its compressive strength. In addition, physical properties such as slump, density, and water absorption are within the acceptable range to be applied as general construction materials. The inclusion of fibers is crucial in order to control crack propagation in geopolymer concrete systems.

The crack propagation is proven to be controlled by the bridging effect, in which the steel fibers function to arrest the formation of macro-cracks that occur from the micro-cracks. This phenomenon led to the improvement of concrete properties, especially flexural strength and toughness. The amount of steel fiber addition is important to ensure the performance of concrete is optimum. Inclusion of 1.5% seems to be the optimum value to ensure the highest performance of geopolymer concrete.

REFERENCES

Al-Majidi, M. H., Lampropoulos, A., & Cundy, A. B. (2017). Steel fiber reinforced geopolymer concrete (SFRGC) with improved microstructure and enhanced fibre-matrix interfacial properties. *Construction and Building Materials, 139,* 286–307.

Al-mashhadani, M. M., Canpolat, O., Aygormez, Y., Uysal, M., & Erdem, S. (2018). Mechanical and microstructural characterization of fiber reinforced fly ash based geopolymer composites. *Construction and Building Materials, 167,* 505–513.

Alomayri, T., Shaikh, F. U. A., & Low, I. M. (2013). Thermal and mechanical properties of cotton fabric-reinforced geopolymer composites. *Journal of Materials Science, 48*(19), 6746–6752.

Aswani, E., & Karthi, L. (2017). A literature review on fiber reinforced geopolymer concrete. *International Journal of Scientific and Engineering Research, 8*(2), 408.

Faris, M. A., Mustafa, M., Bakri, A., Ismail, K. N., Muniandy, R., Kadir, A. A., & Tran, M. (2016). *Review on Different types of Geopolymer Concrete Fibres, 857,* 388–394.

Gao, J., Sun, W., & Morino, K. (1997). Mechanical properties of steel fiber-reinforced, high-strength, lightweight concrete. *Cement and Concrete Composites, 19*(4), 307–313.

Guo, X., & Pan, X. (2018). Mechanical properties and mechanisms of fiber reinforced fly ash-steel slag based geopolymer mortar. *Construction and Building Materials, 179,* 633–641.

Hassanpour, M., Shafigh, P., & Mahmud, H. B. (2014). Mechanical properties of structural lightweight aggregate concrete containing low volume steel fiber. *Arab Journal of Science Engineering, 39,* 3379–3390.

Khan, M. Z. N., Hao, Y., Hao, H., & Shaikh, F. U. A. (2018). Mechanical properties of ambient cured high strength hybrid steel and synthetic fibers reinforced geopolymer composites. *Cement and Concrete Composites, 85,* 133–152.

Korniejenko, K., Frączek, E., Pytlak, E., & Adamski, M. (2016). Mechanical properties of geo-polymer composites reinforced with natural fibers. *Procedia Engineering*, *151*, 388–393.

Lou, X., Xu, J. Y., & Li, W. M. (2009). Research on the basalt fiber reinforced geopolymeric concrete applied to the rapid repair of air field pavement. *Concrete* (in Chinese), *12*, 106–109.

Nematollahi, B., Sanjayan, J., Qiu, J., & Yang, E. H. (2017). High ductile behavior of a poly-ethylene fiber-reinforced one-part geopolymer composite: A micromechanics-based investigation. *Archives of Civil and Mechanical Engineering*, *17*(3), 555–563.

Okoye, F. N., Prakash, S., & Singh, N. B. (2017). Durability of fly ash based geopolymer concrete in the presence of silica fume. *Journal of Cleaner Production*, *149*, 1062–1067.

Ranjbar, N., Talebian, S., Mehrali, M., Kuenzel, C., Cornelis Metselaar, H. S., & Jumaat, M. Z. (2016). Mechanisms of interfacial bond in steel and polypropylene fiber reinforced geopolymer composites. *Composites Science and Technology*, *122*, 73–81.

Reed, M., Lokuge, W., & Karunasena, W. (2014). Fiber-reinforced geopolymer concrete with ambient curing for in situ applications. *Journal of Material Science*, *49*, 4297–4304.

Rendell, F., Jauberthie, R., & Grantham, M. (2002). *Deteriorated Concrete: Inspection and Physicochemical Analysis*. Thomas Telford.

Sanjayan, J. G., Nazari, A., & Pouraliakbar, H. (2015). FEA modelling of fracture toughness of steel fibre-reinforced geopolymer composites. *Materials and Design*, *76*, 215–222.

Sarker, P. K., & McBeath, S. (2015). Fire endurance of steel reinforced fly ash geopolymer concrete elements. *Construction and Building Materials*, *90*, 91–98.

Shafigh, P., Mahmud, H., & Jumaat, M. Z. (2011). Effect of steel fiber on the mechanical properties of oil palm shell lightweight concrete. *Materials & Design*, *32*, 3926–3932.

Zhao, C. F., & Chen, J. Y. (2013). Damage mechanism and mode of square reinforced concrete slab subjected to blast loading. *Theoretical and Applied Fracture Mechanics*, *63–64*, 54–62.

4 Geopolymer Lightweight Aggregate

*Rafiza Abdul Razak, Alida Abdullah, and
Mohd Mustafa Al Bakri Abdullah*
Universiti Malaysia Perlis

Petrica Vizureanu
Gheorghe Asachi Technical University

Eva Arifi
Universitas Brawijaya

Andrei Victor Sandu
Gheorghe Asachi Technical University

4.1 INTRODUCTION

The quality of concrete is strongly influenced by the size, shape, and composition of the aggregates used in the manufacturing process. As previously reported, the type of aggregate affects mainly its strength and density (Naderi & Kaboudan, 2021; Robalo et al., 2021; Schumacher et al., 2020). The choice of aggregates should be based on the desired properties of the concrete. Different aggregates can be used to increase the strength, density, or durability of the concrete. Additionally, the size of the aggregate can also affect the workability of the concrete. Besides that, a significant element in altering concrete's weight is the choice of raw materials used in its manufacturing, especially the aggregates that serve as the structure of the material. More than 70% of the concrete matrix is made up of aggregate, which is the most commonly utilized construction material (Collivignarelli et al., 2020).

The two main categories of lightweight aggregate are those produced by thermal processing from either naturally occurring resources or industrial by-products, and those that are ready for use only after mechanical processing when they occur naturally. Expanded perlite (Ibrahim et al., 2020), vermiculite (Gencel et al., 2021), and natural pumice (Karthika et al., 2020) are a few examples of naturally occurring lightweight aggregates that have been used in producing lightweight concrete. Moreover, lightweight aggregate produced from various industrial by-products, such as expanded glass (Liaver and Poraver) and expanded clay (Liapor and Leca), was used as lightweight aggregate to produce lightweight concrete (Chung et al., 2021). The production of lightweight aggregates from waste may significantly lessen

DOI: 10.1201/9781003390190-4

the environmental impact of rehabilitation and extraction and introduce enormous quantities of different types of waste to the construction market (Lee et al., 2021). Additionally, the effectiveness of lightweight aggregate derived from industrial waste products depends on the various binding agents and production techniques used (Vali & Murugan, 2020).

According to BS EN 13055-1 (2002), the density of lightweight aggregate must be less than 2,000 kg/m^3. Three types of hardening methods are used to produce lightweight aggregate: sintering, cold bonding, and autoclaving. However, sintering is the best method for producing aggregate because it results in microstructure changes at high temperatures (Kamal & Mishra, 2020). The application of sintering to the products by heating the samples in a furnace at a high temperature is one of the industrial procedures for producing lightweight aggregate. The researchers discovered that weak bonding develops below 1,000°C and that complete densification happens at 1,200°C (Ozkan & Kabay, 2022; Sun et al., 2021). Typically, the temperature employed is higher than 1,200°C. Utilizing high temperatures could result in significant energy consumption, which could raise the cost of production (Chinnu et al., 2021; Sun et al., 2021). The lightweight aggregates created by the sintering technique have improved durability features like permeability and corrosion resistance, despite high manufacturing costs (George & Revathi, 2020; Sahoo et al., 2020). There are numerous benefits to using lightweight aggregate in concrete. The lightweight aggregate will reduce costs in other areas, including transportation, labor, and other expenses. In addition, application of lightweight aggregate in concrete can reduce the dead load, give great thermal efficiency, and increase fire resistance (Federowicz et al., 2021; Zhang et al., 2019).

This research has studied the development of lightweight aggregate using fly ash (FA) and volcanic ash (VA) as the geopolymer's precursor materials. In order to compare the production of lightweight geopolymer aggregate utilizing FA and VA at low sintering temperatures (less than 1,000°C), this study will demonstrate the effects of sintering temperature in terms of physical and mechanical properties as well as morphology.

4.2 MIX DESIGN AND PROCESS OF GEOPOLYMER LIGHTWEIGHT AGGREGATE

The class F FA was chosen from the power plant in Manjung, Perak, while the raw VA was gathered in Sidoarjo, Indonesia. An alkaline activator (AA) solution made of sodium hydroxide (NaOH) and sodium silicate (Na$_2$SiO$_3$) is used to activate these raw ingredients. These two solutions are combined and heated until they are homogenous. Based on their ideal design, the VA/AA ratio was set at 1.7, the Na$_2$SiO$_3$/NaOH ratio at 0.6, and the NaOH molarity at 12 M (Razak et al., 2015). In this investigation, the FA/AA ratio, Na$_2$SiO$_3$/NaOH ratio, and NaOH molarity were fixed at 12 M, 3.0, and 2.5 (Abdullah et al., 2021), respectively, to achieve the best mix design for FA. After the mixing procedure, the aggregate was palletized manually to create a sphere shape, and it was then allowed to dry at 60°C. The aggregate was then sintered at various sintering temperatures to compare its characteristics. Figure 4.1 summarizes the steps to create lightweight geopolymer aggregate using FA and VA at various sintering temperatures.

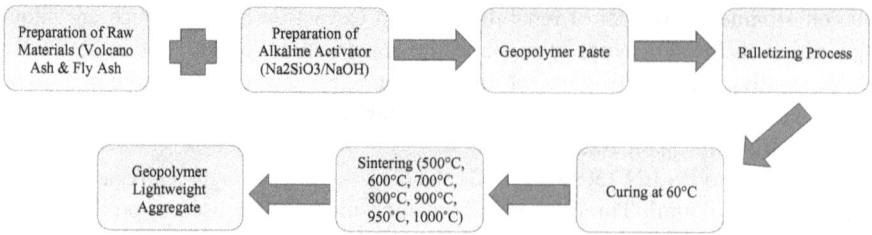

FIGURE 4.1   Flow chart of methodology to produce geopolymer lightweight aggregate.

TABLE 4.1
Chemical Composition of Raw FA and VA

Chemical Composition	Fly Ash (%)	Volcano Ash (%)
SiO_2	55.90	40.00
Al_2O_3	28.10	14.60
Fe_2O_3	6.97	23.25
CaO	3.84	5.46
TiO_2	2.21	1.75
K_2O	1.55	4.28
ZrO_2	0.14	–
V_2O_5	0.09	0.06
MnO	–	0.34
SO_3	–	0.88
LOI	1.20	9.38

4.3   CHARACTERIZATION ANALYSIS OF RAW MATERIALS

4.3.1   CHEMICAL COMPOSITION ANALYSIS

The chemical composition of FA and VA has been determined using an X-ray fluorescence (XRF) test. Table 4.1 shows the results for the chemical composition of FA and VA. For this FA, the total composition of $SiO_2 + Al_2O_3 + Fe_2O_3$ is 90.97%, which is higher than 70%, indicating that this FA can be classified as class F FA according to ASTM C 618-12a and is suitable to be used as a raw material for geopolymer, while the chemical composition for VA shows that the total of $SiO_2 + Al_2O_3 + Fe_2O_3$ is 77.85%, which is higher than 70%, indicating that this VA can be used as pozzolan materials according to ASTM C618-92a. From the results, this VA is suitable to be used as one of the source materials for geopolymerization.

The results from chemical composition for both FA and VA show that the highest content in these raw materials is Si, with 55.90% for FA and 40.00% for VA. In the geopolymerization process, there will be more silicate species available for the reaction if the Si content in the source materials is high. The Si content is responsible for the durability and strength development in geopolymer systems.

The second highest content of the element in FA is Al_2O_3, which consists of 28.10%, and the second highest element in VA is Fe_2O_3, with 23.25%. A high ratio of

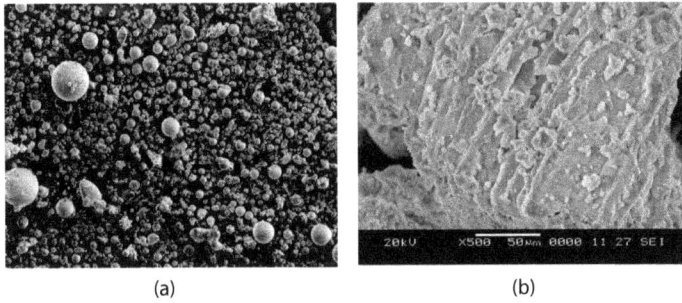

(a) (b)

FIGURE 4.2 Microstructure image of (a) FA at 500× magnification and (b) VA at 500× magnification.

SiO_2/Al_2O_3 led to a high setting time, which resulted in a higher strength. In this FA, SiO_2/Al_2O_3 is 1.99, which is very close to the suggested ratio by Davidovits (1999), which is 2.0, and for VA, the ratio of SiO_2/Al_2O_3 is 1.72. In the geopolymerization process, Al is important for setting time for the formation of geopolymer. The Al component tends to dissolve more easily than the silicon component during geopolymerization because its atomic number is lower than that of Si.

4.3.2 MORPHOLOGY ANALYSIS

The morphology analysis of FA is presented in Figure 4.2a, which clearly shows that FA comprises different sizes of spherical vitreous particles and different shapes of particles. These spherical particles are hollow, and some spheres may contain other finer, smaller particles in their interior bodies. Figure 4.2b shows the microstructure image of VA at 500×. It has a plate-like structure or flaked-shaped particles, which is close to clay structure.

From the morphology analysis results, it clearly shows that the particle size of FA is smaller than VA, with less than 10 μm. Meanwhile, the overall particle size of VA is dominated by particles between 2.5 and 25.0 μm. The spherical shape of FA particles will help to improve the workability in the mixing of aggregates, whereas the small particle size plays a role as a filler for voids, therefore producing dense and durable concrete.

4.4 EFFECT OF SINTERING TEMPERATURE TO THE PROPERTIES OF GEOPOLYMER LIGHTWEIGHT AGGREGATE

4.4.1 PHYSICAL PROPERTIES

Density. The density is a critical factor in determining whether the resulting aggregate can be categorized as lightweight. Figure 4.3 displays the density of a geopolymer lightweight aggregate made from FA and VA produced at various sintering temperatures. The graph demonstrates that a lower density value is produced at a higher sintering temperature for both samples. VA geopolymer lightweight aggregate density is produced at 1,950 kg/m³ at 500°C and only drops to 1,650 kg/m³ at

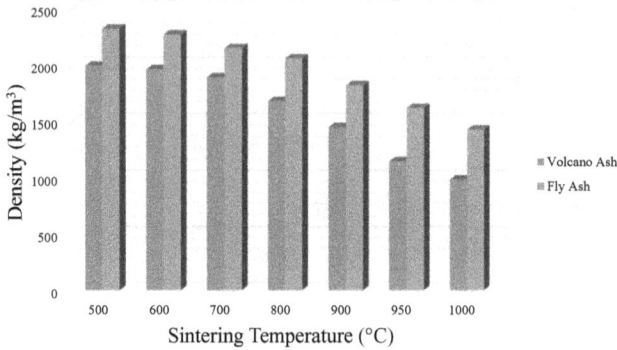

FIGURE 4.3 Graph of density for geopolymer lightweight aggregate at various sintering temperatures.

800°C. The density is 1,120 kg/m³ at 950°C for sintering, and it drops to 980 kg/m³ at 1,000°C, which is less than 1,000 kg/m³.

FA is used as a lightweight aggregate in geopolymers, and its density exhibits the same pattern as that of VA, with decreasing density values at higher sintering temperatures. The density measured is 2,340 kg/m³ at 400°C for sintering. Density continuously drops until 700°C, when it is 2,050 kg/m³. Since lightweight aggregate is defined as having a density of less than 2,000 kg/m³ and starting at 800°C, the density reported is 1,820 kg/m³, which marks the beginning of the process. As the temperature of the sintering process rises to 1,000°C, the aggregate density value decreases. Larger pores that were present at 1,000°C may have affected the aggregate density formed for both VA and FA samples. When the sintering temperature is increasing, the aggregate density mainly depends on the type of materials used. The FA used in this study had a higher density than VA, which caused the density of the lightweight aggregate to become higher. All figures fall under the category of lightweight aggregate because they are less than 2,000 kg/m³ (BS EN 13055-1, 2002).

Water Absorption. Figure 4.4 demonstrates that FA as a lightweight aggregate has a lower water absorption rate than VA. At 700°C and 500°C, the VA and FA geopolymer lightweight aggregates exhibit maximum water absorption rates of 11.3% and 6.33%, respectively. At lower sintering temperatures, the pellet became looser, and the creation of the vitrified exterior layer was incomplete. The pellets will have excessive open pores if the sintering temperature is too low, which makes them unable to produce the crystalline phase between the materials' particles (Lim et al., 2019). However, VA and FA geopolymer lightweight aggregates exhibit the lowest water absorption, with values of 5.30% (at 950°C) and 1.52% (at 900°C). The graph unmistakably demonstrates that FA absorbs more water than VA for all samples, demonstrating that FA geopolymer lightweight aggregate is denser than VA, as demonstrated by the density result. Due to the porous inner and outer surfaces of the generated geopolymer lightweight aggregate, lower sintering temperatures nevertheless result in considerable water absorption. As a result, the porous structure's propensity to absorb

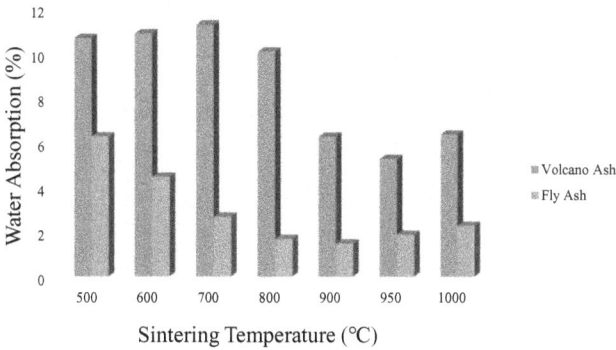

FIGURE 4.4 Graph of water absorption for geopolymer lightweight aggregate at various sintering temperatures.

water contributes to its high density. The creation of a dense vitrified shell, which can inhibit the high absorption of water, causes the lowest water absorption to be observed at 900°C and 950°C for FA and VA samples, respectively (Kwek et al., 2022; Ren et al., 2020). Furthermore, the surface of the aggregate is smooth and shows very few pores in the lightweight aggregate's surrounding area at these ideal sintering temperatures. However, at 1,000°C, more pores were created at the lightweight aggregate's surface. The open porosity that appeared on the surface of the sintered samples correlated with overburning and expansion of the samples, which increased water absorption (Zeng et al., 2019). At the same time, larger pores were noticed at the center of the aggregate, which increased water absorption and contributed to the lowest density.

4.4.2 MECHANICAL PROPERTIES

This section describes the aggregate impact value (AIV) for artificial lightweight geopolymer aggregate manufactured using VA and FA at different sintering temperatures.

4.4.2.1 Aggregate Impact Value

A percentage of AIV with a low value implies that the strength of aggregate in concrete is exceptional. In accordance with BS 812-112 (1990), the AIV should not exceed 30%. Figure 4.5 depicts the AIV of a lightweight geopolymer aggregate manufactured using VA and FA at various sintered temperatures.

The lowest AIV for VA geopolymer lightweight aggregate that contributes to the highest strength is at 950°C with 15.79%. The lowest AIV value for FA geopolymer lightweight aggregate is recorded at 900°C for sintering, with 14.5%. Overall, FA geopolymer lightweight aggregate demonstrates more strength than VA geopolymer lightweight aggregate. In contrast to VA, which requires 950°C to attain the best strength of aggregate produced, FA has an advantage since the best strength, as demonstrated by AIV results at 900°C, can be obtained at 900°C, which can save more energy.

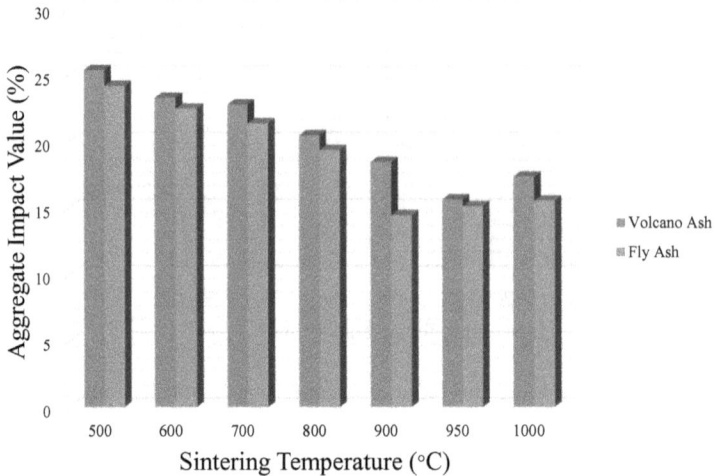

FIGURE 4.5 Graph of AIV for geopolymer lightweight aggregate at various sintering temperatures.

However, the density of these aggregates varied and impacted the aggregate's strength. The increased density of FA geopolymer lightweight aggregate compared to that of VA geopolymer lightweight aggregate contributed to FA geopolymer lightweight aggregate's better strength.

The fusing of aluminum-silicate minerals can form stronger connections during the sintering process. The underlying microstructural change causes the bonds to get stronger and become harder when the sintering temperature is increased. Besides that, the development of a vitrified shell, which strengthens the structure of the lightweight aggregate made of geopolymer, is the cause of these high strengths for both samples (Kwek et al., 2022). However, from 500°C to 1,000°C, the total value of AIV for the two samples of VA and FA shows a little difference. Low strength was caused by the significant water absorption at lower sintering temperatures (500°C–800°C), as seen by the high AIV value, which varied from 19.50% to 25.44%. As the sintering temperature increased, the pellet revealed a rigid and disconnected pore structure. However, when the temperature was above optimal, the aggregate structure was less dense and less vitrified (Lim et al., 2019). Due to larger pores and increased water absorption caused by the production of geopolymer lightweight aggregate at 1,000°C, the AIV value is rising and weakening the structure's strength. Moreover, the pores in the microstructure of lightweight aggregate changed shape after the optimal sintering temperature, transitioning from smaller pores to pores with a larger size. Although there are smaller pores at the optimum sintering temperature, it still provides sufficient strength to the aggregate (Li et al., 2021). From the overall result, both VA and FA geopolymer lightweights had an AIV value less than 30%, which can be used as material in the concrete (BS 812-112, 1990).

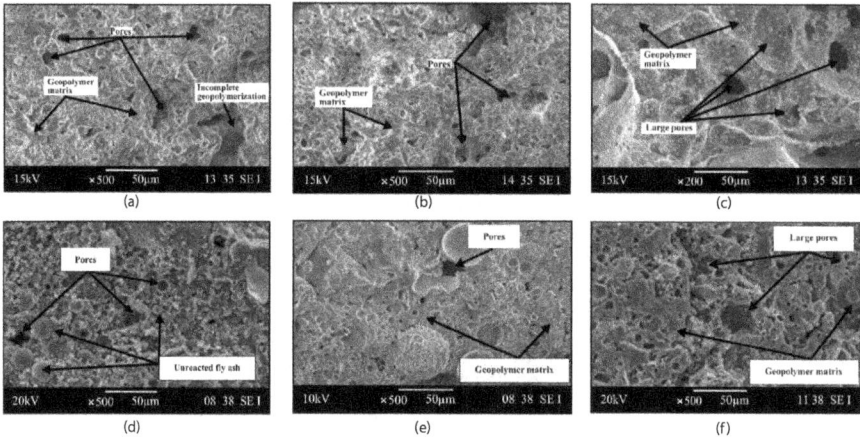

FIGURE 4.6 SEM images of geopolymer lightweight aggregate using volcanic ash (VA) produced at (a) 900°C, (b) 950°C, (c) 1,000°C; geopolymer lightweight aggregate using fly ash (FA) produced at (d) 900°C, (e) 950°C, and (f) 1,000°C.

4.4.3 MORPHOLOGY OF GEOPOLYMER LIGHTWEIGHT AGGREGATE

Figure 4.6 illustrates the SEM pictures of a geopolymer lightweight aggregate made with VA and FA and sintered at 900°C, 950°C, and 1,000°C. Every sintering temperature sample has pores that are present. The lightweight aggregate has pores with diameters ranging from 3.19 to 20.08 μm when sintered at 900°C and 950°C. For VA samples, the distribution of pores is relatively uniform, increasing the amount of geopolymer matrix formed and contributing to the aggregate's high strength (Fu et al., 2021) (Figure 4.6b). Larger pores with pore sizes ranging from 25 to 140 μm are visible at the 1,000°C sintering temperature (Figure 4.6c), impacting water absorption and decreasing strength. In the meantime, FA samples have a denser matrix that contributes to the lightweight geopolymer FA aggregate's excellent strength (Figure 4.6d). However, for VA and FA geopolymer aggregates, several big pores can be seen at 1,000°C, contributing to a loss of strength. In addition, the larger pore caused an increase in water absorption, but it also reduced the density of the aggregate. Moreover, microstructure of VA geopolymer aggregate at 900°C proved that incomplete geopolymerization causes higher water absorption and lower strength as compared to the rigid geopolymer matrix observed at 950°C. For FA geopolymer aggregate, the unreacted FA observed at 900°C did not affect aggregate properties, consisting of lower water absorption and higher strength.

4.5 CONCLUSION

The AIV and density of geopolymer lightweight aggregate were found to be significantly affected by the sintering temperature. From the study, the following conclusions were drawn.

Research on geopolymer lightweight aggregate utilizing VA revealed that the ideal AIV is 15.79% and that a low density of 1,120 kg/m^3 and water absorption are achieved at 950°C during sintering (5.3%).

Research employing FA for geopolymer lightweight aggregate revealed that the ideal AIV value is 14.5%, with a density of 1,820 kg/m^3 and a water absorption of 1.52% at 900°C during sintering.

Both the VA and FA lightweight geopolymers had an AIV value of less than 30%, allowing them to be employed as lightweight aggregates in concrete.

REFERENCES

Abdullah, A., Hussin, K., Abdullah, M. M. A. B., Yahya, Z., Sochacki, W., Razak, R. A., Bloch, K., & Fansuri, H. (2021). The effects of various concentrations of NaOH on the inter-particle gelation of a fly ash geopolymer aggregate. Materials, 14(5), 1–11.

ASTM C 618-12a. (2012). *Standard Specification for Coal Fly Ash and Raw or Calcined Natural Pozzolan for use in Concrete*. ASTM International, West Conshohocken, PA, USA.

ASTM C618-92a. (1994). *Standard Specification for Fly Ash and Raw or Calcinated Natural Pozzoland for Use as Mineral Admixture in Portland Cement Concrete: Annual Book of ASTM Standards*. American Standard for Testing Materials, Pennsylvania

BS EN 13055-1 British Standard. (2002). *Lightweight Aggregate—Part 1: Lightweight Aggregates for Concrete, Mortar and Grout*. British Standard.

BS 812-112 British Standard. (1990). *Testing Aggregates—Part 112: Methods for Determination of Aggregate Impact Value (AIV)*. British Standard.

Chinnu, S. N., Minnu, S. N., Bahurudeen, A., & Senthilkumar, R. (2021). Recycling of industrial and agricultural wastes as alternative coarse aggregates: A step towards cleaner production of concrete. *Construction and Building Materials*, *287*, 123056.

Chung, S. Y., Sikora, P., Kim, D. J., El Madawy, M. E., & Abd Elrahman, M. (2021). Effect of different expanded aggregates on durability-related characteristics of lightweight aggregate concrete. *Materials Characterization*, *173*, 110907.

Collivignarelli, M. C., Cillari, G., Ricciardi, P., Miino, M. C., Torretta, V., Rada, E. C., & Abbà, A. (2020). The production of sustainable concrete with the use of alternative aggregates: A review. *Sustainability* (Switzerland), *12*(19), 1–34.

Davidovits, J. (1999). Chemistry of geopolymeric systems. In: *Geopolymer '99 International Conference*, France, pp. 9–40.

Federowicz, K., Techman, M., Sanytsky, M., & Sikora, P. (2021). Modification of lightweight aggregate concretes with silica nanoparticles-a review. *Materials*, *14*(15). doi:10.3390/ma14154242

Fu, Q., Xu, W., Zhao, X., Bu, M. X., Yuan, Q., & Niu, D. (2021). The microstructure and durability of fly ash-based geopolymer concrete: A review. *Ceramics International*, *47*(21), 29550–29566.

Gencel, O., Gholampour, A., Tokay, H., & Ozbakkaloglu, T. (2021). Replacement of natural sand with expanded vermiculite in fly ash-based geopolymer mortars. *Applied Sciences (Switzerland)*, *11*(4), 1–18.

George, G. K., & Revathi, P. (2020). Production and utilisation of artificial coarse aggregate in concrete: A review. *IOP Conference Series: Materials Science and Engineering*, *936*(1), 12035.

Ibrahim, M., Ahmad, A., Barry, M. S., Alhems, L. M., & Mohamed Suhoothi, A. C. (2020). Durability of structural lightweight concrete containing expanded perlite aggregate. *International Journal of Concrete Structures and Materials*, *14*(1), 1–15.

Kamal, J., & Mishra, U. K. (2020). Influence of fly ash properties on characteristics of manufactured angular fly ash aggregates. *Journal of the Institution of Engineers (India): Series A*, *101*(4), 735–742.

Karthika, R. B., Vidyapriya, V., Nandhini Sri, K. V., Merlin Grace Beaula, K., Harini, R., & Sriram, M. (2020). Experimental study on lightweight concrete using pumice aggregate. *Materials Today: Proceedings, 43*, 1606–1613.

Kwek, S. Y., Awang, H., Cheah, C. B., & Mohamad, H. (2022). Development of sintered aggregate derived from POFA and silt for lightweight concrete. *Journal of Building Engineering, 49*, 104039.

Lee, K. H., Lee, K. G., Lee, Y. S., & Wie, Y. M. (2021). Manufacturing and application of artificial lightweight aggregate from water treatment sludge. *Journal of Cleaner Production, 307*, 127260.

Li, X., He, C., Lv, Y., Jian, S., Jiang, W., Jiang, D., Wu, K., & Dan, J. (2021). Effect of sintering temperature and dwelling time on the characteristics of lightweight aggregate produced from sewage sludge and waste glass powder. *Ceramics International, 47*(23), 33435–33443.

Lim, Y. C., Lin, S. K., Ju, Y. R., Wu, C. H., Lin, Y. L., Chen, C. W., & Dong, C. Di. (2019). Reutilization of dredged harbor sediment and steel slag by sintering as lightweight aggregate. *Process Safety and Environmental Protection, 126*, 287–296.

Naderi, M., & Kaboudan, A. (2021). Experimental study of the effect of aggregate type on concrete strength and permeability. *Journal of Building Engineering, 37*, 101928.

Ozkan, H., & Kabay, N. (2022). Manufacture of sintered aggregate using washing aggregate sludge and ground granulated blast furnace slag: Characterization of the aggregate and effects on concrete properties. *Construction and Building Materials, 342*(PB), 128025.

Razak, R. A., Abdullah, M. M. A. B., Hussin, K., Ismail, K. N., Hardjito, D., & Yahya, Z. (2015). Optimization of NaOH molarity, LUSI mud/alkaline activator, and Na_2SiO_3/NaOH ratio to produce lightweight aggregate-based geopolymer. *International Journal of Molecular Sciences, 16*(5), 11629–11647.

Ren, Y., Ren, Q., Huo, Z., Wu, X., Zheng, J., & Hai, O. (2020). Preparation of glass shell fly ash-clay based lightweight aggregate with low water absorption by using sodium carbonate solution as binder. *Materials Chemistry and Physics, 256*, 123606.

Robalo, K., Costa, H., do Carmo, R., & Júlio, E. (2021). Experimental development of low cement content and recycled construction and demolition waste aggregates concrete. *Construction and Building Materials, 273*, 121680.

Sahoo, S., Selvaraju, A. K., & Suriya Prakash, S. (2020). Mechanical characterization of structural lightweight aggregate concrete made with sintered fly ash aggregates and synthetic fibres. *Cement and Concrete Composites, 113*, 103712.

Schumacher, K., Saßmannshausen, N., Pritzel, C., & Trettin, R. (2020). Lightweight aggregate concrete with an open structure and a porous matrix with an improved ratio of compressive strength to dry density. *Construction and Building Materials, 264*, 120167.

Sun, Y., Li, J. S., Chen, Z., Xue, Q., Sun, Q., Zhou, Y., Chen, X., Liu, L., & Poon, C. S. (2021). Production of lightweight aggregate ceramsite from red mud and municipal solid waste incineration bottom ash: Mechanism and optimization. *Construction and Building Materials, 287*, 122993.

Vali, K., & Murugan, S. (2020). Influence of industrial by-products in artificial lightweight aggregate concrete: An environmental benefit approach. *Ecology, Environment and Conservation, 26*, S233–S241.

Zeng, L., Sun, H. J., Peng, T. J., & Zheng, W. M. (2019). The sintering kinetics and properties of sintered glass-ceramics from coal fly ash of different particle size. *Results in Physics, 15*, 102774.

Zhang, J., Ma, G., Huang, Y., sun, J., Aslani, F., & Nener, B. (2019). Modelling uniaxial compressive strength of lightweight self-compacting concrete using random forest regression. *Construction and Building Materials, 210*, 713–719.

5 Lightweight Geopolymer-Based Ceramics

Romisuhani Ahmad, Mohd Mustafa Al Bakri Abdullah, and Wan Mastura Wan Ibrahim
Universiti Malaysia Perlis

Andrei Victor Sandu
Gheorghe Asachi Technical University

5.1 INTRODUCTION

Lightweight ceramic can be utilized in a variety of applications, including construction, cutting tools, amour systems, wear lining refractories, and wastewater treatment, depending on its density. Traditionally, ceramics has been defined as an inorganic, nonmetallic material consisting of metallic and nonmetallic elements bonded together with ionic and/or covalent bonds. Ceramics that can be classified as traditional and advanced ceramics can be polycrystalline or at least partly polycrystalline structures formed by the sintering process (Grigoriev et al., 2019). The applications for traditional ceramics and glasses include structural building materials, refractories for furnace linings, sanitaryware and tableware, transportation, and electrical insulation (Mukherjee, 2013). Whereas a variety of applications for advanced ceramics have been developed with the intention of expanding at a reasonable rate where the processing tolerance and cost effectiveness compare to traditional ceramics.

The superior character of ceramic materials contributes to their huge application in this modern industry. Generally, ceramics were labeled as refractory due to their high melting points. Different types of ceramics have various properties; in general, they are also high in modulus, high compression strength, high hardness, low thermal conductivity, and chemically inert (Popoola et al., 2014). The ionic and covalent bonds in the crystal structure of ceramic materials influence the stability of bonding. Although the strong bonds increase the fragility of the ceramics, which in some way limits the applications (Iyasara et al., 2014). Additionally, the industry is very interested in the recent rise in demand for ceramic materials that are stiffer, stronger, and lighter. Thus, the selected materials used, suitable fabrication methods, and sintering can all affect the morphology of the materials to meet the required properties and performance.

Few methods are available for fabricating ceramic materials, and the fabrication method is very important since it affects the properties of the product. Ceramic

DOI: 10.1201/9781003390190-5

fabrication is accomplished through a series of steps that begin with raw material, progress through batch preparation and forming, and conclude with firing. Fabrication of conventional ceramic materials necessitates extremely high temperatures up to 1,600°C (Kriven et al., 2010). The geopolymer method is an alternate approach for fabricating ceramic materials since the amorphous to semi-crystalline nature of geopolymers will change into crystalline ceramic phases when heated. With the benefit of geopolymerization reactions, high-temperature techniques are no longer necessary to produce ceramic materials structure and properties (Iwahiro et al., 2001). Additionally, geopolymer can be used to practically design the chemical compositions of the final product and be immediately changed into the desired structural ceramic component (Zulkifli et al., 2020).

Geopolymer is an inorganic aluminosilicate produced at low temperatures that has potential applications as a matrix in ceramics, coatings, cement, and other composite materials (Wan Ibrahim et al., 2020; Shahari et al., 2021; Kovářík et al., 2021). In general, geopolymers are produced by mixing the high content of aluminosilicate source material with an alkali solution, and the mixture is cured at a certain temperature (Cong and Cheng, 2021). Sodium hydroxide (NaOH) and potassium hydroxide (KOH) used as alkali solutions will incorporate into source materials rich in SiO_2 and Al_2O_3, yielding Si–O–Al–O bonds. Any alkali and alkali earth cations can be used as the alkali component. However, geopolymerization is a multiphase reaction that consists of a dissolution-reorientation-solidification series (Davidovits and Quentin, 1991; Duxson et al., 2005).

The nature, chemical composition, reactivity, and fineness of the source materials were discovered to affect the strength of geopolymers. Geopolymer synthesized using calcined source materials, e.g., metakaolin, slag, and fly ash, has a higher compressive strength compared to those synthesized using non-calcined materials, such as kaolinite, and naturally occurring minerals (Palomo et al., 1999; Barbosa and MacKenzie, 2003; Xu and Van Deventer, 2000). Various studies on geopolymer ceramics focused on the use of metakaolin (Bell et al., 2009; Peigang et al., 2011; Peigang et al., 2013) due to its high reactivity, which is slightly more reactive than the less reactive kaolin (Heah et al., 2011).

Despite this, the use of kaolin as a major raw material in the fabrication of ceramic materials has the advantage of being readily available and inexpensive. Since the properties of ceramics are highly dependent on powder packing and additives, a proper additive selection is important in the pursuit of lightweight ceramic materials to enhance the mechanical properties during and after the manufacturing process (Pelz et al., 2021; Rajeswari et al., 2015; Taktak et al., 2011). Common types of additives used in ceramic processing consist of binders, plasticizers, surfactants, dispersants, and lubricants. This chapter presents the design and performance of kaolin geopolymer ceramics (KGC) and KGC with the addition of ultra-high-molecular-weight polyethylene (UHMWPE) as a binder, denoted as KGCB.

5.2 FABRICATION OF LIGHTWEIGHT GEOPOLYMER CERAMICS

The challenge of fabricating ceramic materials using geopolymer for lightweight applications with the excellent properties and performance required can be overcome on a continuous and cost-effective basis. The influence of component composition as well as size, shape, and dimensional-surface finish requirements on fabrication routes and the parameters of the processing technique used within the routes on both goals. Previous research has mostly concentrated on the microstructural and physical evolution of geopolymers after heating. The monolithic bodies often fracture when heated as-cast due to strong capillary forces operating on capillary channels during drying (Ahmad et al., 2020). Peigang et al. (2013) produced a geopolymer-derived leucite ceramic with low mechanical strength.

This defect would retard the practical application of geopolymer technology in advanced ceramics. The solution to this issue is to sinter geopolymer powder that has been crushed and compressed. By employing this technique, the researchers discovered that it offers a straightforward, affordable alternative to the traditional way of producing reasonably high-strength ceramics. The capillary stresses caused by the evaporation of water prevented cracking on heating, thus resulting in better strength. Therefore, the fabrication process is crucial to fabricating ceramics with the greatest mechanical properties and relative densities.

5.3   SINTERING OF LIGHTWEIGHT GEOPOLYMER CERAMICS

Other than the fabrication method, the sintering condition also plays an important role in fabricating ceramic materials with the desired microstructure. Sintering is a technique involving heat treatment that shapes the powder or porous material into the desired shape and turns it into a usable solid (Sames et al., 2016). Furthermore, the sintering temperature has a significant impact on the phase transition of geopolymer from amorphous to crystalline. Xie et al. (2010) reported that the geopolymer was sufficiently sintered over 1,000°C to solidify the sample and begin the leucite crystallization. Maximum values were reached for density, fracture toughness, and biaxial flexure strength after 3 hours of heating at 1,200°C. Toughness and biaxial flexure were influenced by grain size and leucite concentration.

The higher sintering temperature was reported to produce the necessary size and quantity of leucite grains to reinforce the glassy matrix with the stronger, high-temperature, leucite ceramic phase. Peigang et al. (2010) discovered that the crystallization peak temperature of $K_2O \cdot Al_2O_3 \cdot 5SiO_2$ was 986.3°C and the temperature range of sintering was 700°C–954.3°C. Another representative work was done by He et al. (2015), in which a substantial amount of leucite was developed by sintering the geopolymer at 800°C. At a heating rate of 2°C/min, the bullet-shaped and plate-shaped leucite ceramics were observed without cracking by directly sintering the geopolymer precursors.

Generally, high-temperature exposure led to sintering processes, structural rearrangement, and formation of crystalline phases such as nepheline, kalsilite, and mullite. The densification of materials will be improved by the sintering profile, while diffusion functions as a mechanism to favor both densification and grain growth (Hu et al., 2017). Sintering and crystallization can take place in either a sequential or concurrent manner. If crystallization occurs before the sintering process is completed, the viscosity quickly increases and the sintering process ceases, leaving a porous material (Prado et al., 2003). In some cases, the sintering process can occasionally be faster than the crystallization process, depending on the heating rate.

5.4 EFFECT OF SINTERING TEMPERATURE

Heat treatment has a major impact on the mechanical strength of the geopolymer ceramics as water evaporates during sintering. With the optimum concentration of 12 M NaOH and Na_2SiO_3/NaOH ratios of 0.24 in producing kaolin as the source material of geopolymer, the effect of sintering temperature on KGC and KGC with the addition of 4 wt.% UHMWPE as binder (KGCB) was studied.

5.4.1 FLEXURAL STRENGTH ANALYSIS

Figure 5.1 shows the flexural strength of KGC and KGCB at different sintering temperatures. The results of both samples demonstrate that flexural strength increases as the sintering temperature rises. However, compared to samples without the addition of UHMWPE, the addition of UHMWPE results in increased flexural strength at all sintering temperatures. Sintering the KGC to 900°C results in the lowest flexural strength of 37.1 MPa, while the highest flexural strength of 92.1 MPa is obtained by KGCB at the highest sintering temperature of 1,200°C. The effect of UHMWPE as a binder on KGC is recognizable with the increasing strength of the KGC at all sintering temperatures. The samples sintered at 900°C had the lowest flexural strength due to the low temperature, which was not sufficient to cause the particles to diffuse. This is due to the possibility that high temperatures could make the system denser, which would force the liquid phase to fill the gaps and occupy the pores. As a result, there was a low point of contact between the particles, which led to the development of open and interconnected pores. The highest flexural strength achieved at the sintering temperature of 1,200°C indicates that the sample particle was evenly distributed, resulting in a large sintered area and therefore, forming a smooth matrix.

5.4.2 DENSITY ANALYSIS

Figure 5.2 shows the density of KGC and KGBC at different sintering temperatures. The density measured decreased with increasing sintering temperatures for both KGC and KGCB. KGC sintered at 900°C has the highest density of 2.67 g/cm³,

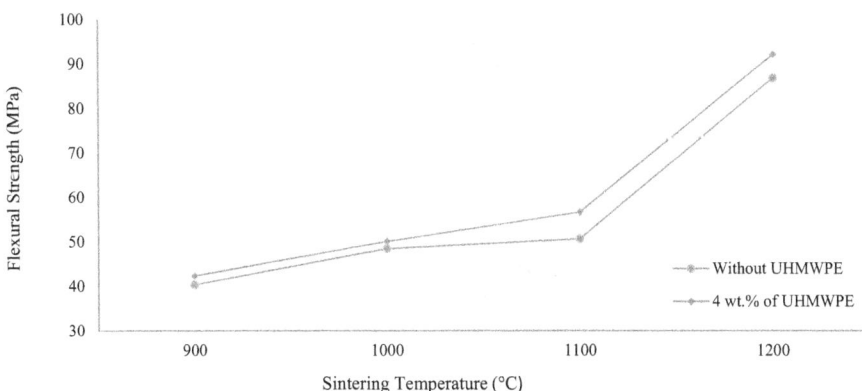

FIGURE 5.1 Flexural strength of KGC and KGCB at different sintering temperatures.

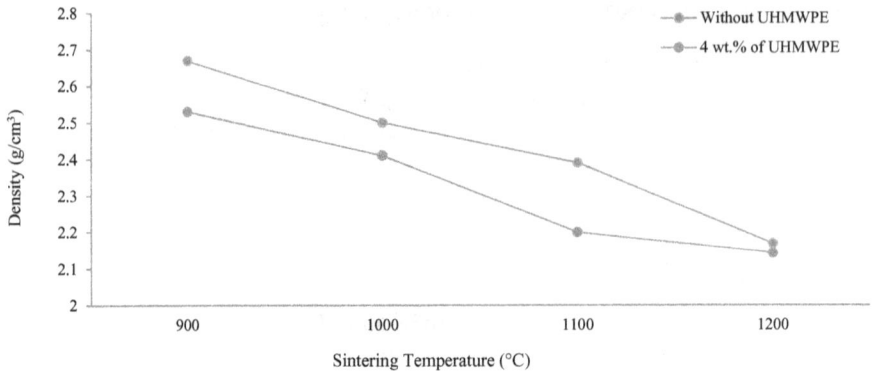

FIGURE 5.2 Density of KGC and KGCB at different sintering temperatures.

decreases to 2.50 g/cm³ at 1,000°C, and gradually decreases to 2.13 g/cm³ at 1,200°C. A decreasing density in the range of 2%–19% was observed on KGCB compared to KGC. Though the same trend was still observed. The lowest density was achieved by the samples with the addition of UHMWPE that was sintered at 1,200°C, at 1.88 g/cm³. The densities of KGC and KGCB decreased with increasing sintering temperatures due to insufficient removal of the entrapped oxygen during sintering. The open porosity on the surface of the samples was closed during sintering, thus preventing the entrapped oxygen from escaping. After the sintering process, the oxygen is left within the green body as porosity. The same trend was found by Olevsky and Molinari (2000), who identified that the increasing sintering temperature causes the formation of pores. Additionally, the structure of the material changes at high temperatures, which is attributed to the liquid sintering. Liquid sintering leads to the closure of previously accessible pores and pore channels (Zawrah et al., 2022), and the decomposition of UHMWPE in the samples also leads to the formation of pores, consequently lowering the density.

5.4.3 WATER ABSORPTION ANALYSIS

The water absorption results of KGC and KGCB at different sintering temperatures are shown in Figure 5.3. The results show a decreasing pattern with an increase in sintering temperature for samples both with and without the addition of UHMWPE. The percentage of water absorption for the KGCB is higher compared to the sample KGC. The result of water absorption at 1,200°C shows the lowest percentage for both samples with and without the addition of UHMWPE, with 6.32% and 1.20%, respectively. KGCB shows a high percentage of water absorption due to the porous structure that is produced during the decomposition of UHMWPE at high temperatures. Micro-cavities exist between particles of the green body of unsintered ceramic materials, and these interparticle spaces are eliminated during the sintering process. The elimination process will result in a reduction of porosity, lowering the water absorption (Karamanov et al., 2009). A similar trend

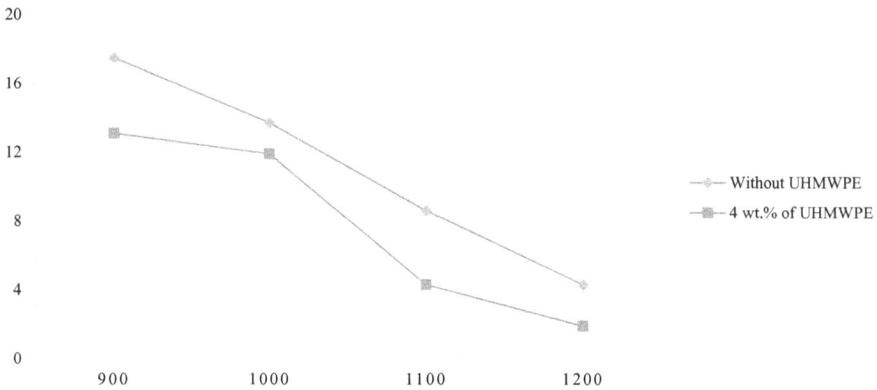

FIGURE 5.3   Water absorption of KGC and KGCB at different sintering temperatures.

was discovered by Baccour et al. (2009): water absorption is closely related to densification. A more substantial liquid phase formation at high temperatures penetrated the pores, closing them as well as isolating the neighboring pores.

5.4.4   PHASE ANALYSIS

XRD patterns of kaolin geopolymer and KGCB at different sintering temperatures are shown in Figure 5.4. The crystallographic compositions of the kaolin geopolymer changed as the broad hump of the amorphous phase disappeared when the kaolin geopolymer underwent heat treatment. All of the samples sintered in the range 900°C to 1,200°C show very similar peaks, with the appearance of the same nepheline ($NaAlSiO_4$) and carbon (C) peaks. However, the intensity of the nepheline and carbon peaks increased with the increasing sintering temperature of the kaolin geopolymer. From the XRD pattern, it can be observed that the disappearance of a broad hump in kaolin geopolymer resulted from the phase transformation from the amorphous phase to the crystalline phase during sintering. As identified by Baccour et al. (2009), two kinds of processes take place during sintering: decomposition and phase transformations. A new phase corresponding to nepheline was detected due to the use of a sodium-based activator in geopolymer synthesis. The intensity of the peaks increases with the increase in sintering temperature, indicating an increase in crystallinity and consequently improving the mechanical strength of the product, as discussed in Section 4.4.1. Crystallization of nepheline from kaolin geopolymer occurred steadily, and this was agreed upon by a few researchers (Kuenzel et al., 2013; Marković et al., 2006) who detected nepheline after exposure to 900°C.

5.4.5   MORPHOLOGY ANALYSIS

The SEM micrographs of KGC and KGCB at different sintering temperatures of 900°C, 1,000°C, 1,100°C, and 1,200°C are shown in Figures 5.5 and 5.6, respectively. It is observed that the SEM micrograph of the sample sintered at 900°C

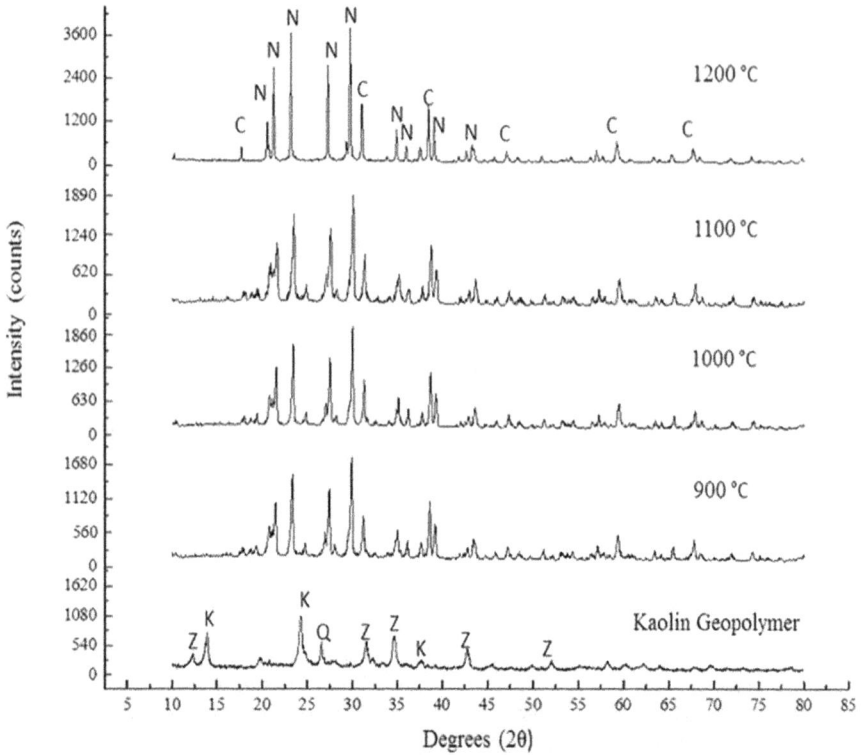

FIGURE 5.4 XRD pattern of kaolin geopolymer and KGCB at different sintering temperatures (N=nephaline, C=carbon, K=kaolinite, Q=quartz, Z= zeolite).

and 1,000°C shows a sponge-like gel texture. As the samples sintered to a higher temperature of 1,100°C and 1,200°C, a relatively well-developed microstructure produced a more compacted and smoother surface. Obvious open and closed pores were spotted in all the SEM micrographs of the KGC with different sintering temperature samples. Despite this, the increasing sintering temperature demonstrated well-shaped and distributed pores compared to the low sintering temperature. The SEM micrograph clearly shows a significant change in the structure of the KGCB, with the increasing sintering temperature simultaneously giving a higher value of strength. The higher sintering temperature sustained the consolidation and aided in a fairly uniform microstructure, thus creating a smooth surface texture. Reported that sintering the geopolymer consequently resulted in the formation of a liquid phase, which allows the joining of particles and the transformation of a plate-like structure into a dense microstructure. The presence of small pores throughout the sample denotes the transformation of amorphous to nepheline crystalline ceramic. It is known that sintering leads to phase transformation (Baccoura et al., 2009). Furthermore, the addition of UHMWPE to the kaolin geopolymer also created

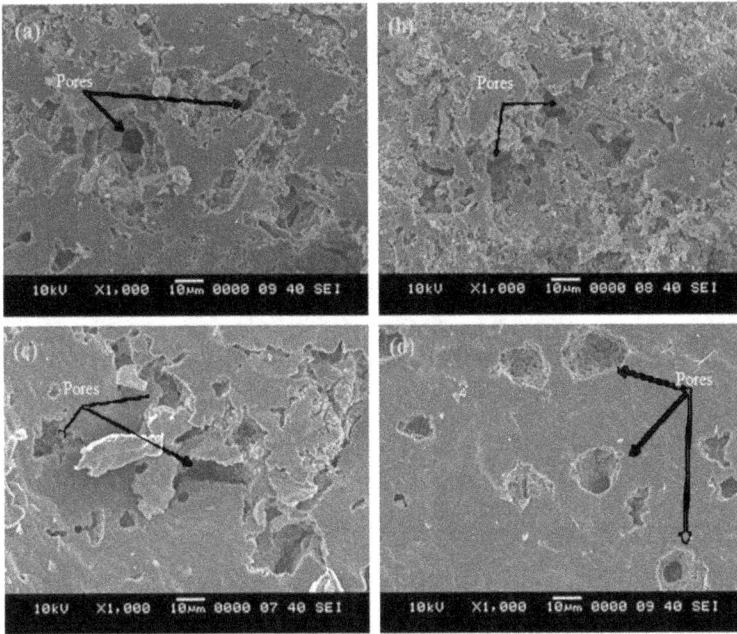

FIGURE 5.5 SEM micrograph of KGC with different sintering temperatures (a) 900°C, (b) 1,000°C, (c) 1,100°C, and (d) 1,200°C.

pores through the decomposition process, as mentioned before. This finding is supported by the decrease in density as the sintering temperature increased.

5.5 CONCLUSION

From the present experimental data, the feasibility of the production of geopolymer ceramic with lightweight properties could be highly achieved if it is combined with other binders. Including geopolymer technology as an alternate substituent in the fabrication of ceramic materials leads to the extent of its applications. In 1970, a new orientation of alkali activation or geopolymerization process were introduced as a new technology in this research area. Even so, it has faced many difficulties in terms of acceptance by the research committees. The use of metakaolin rather than kaolin in research investigations has been considered the main interest of many researchers, owing to its higher reactivity compared to kaolin. This could be explained by the calcination of metakaolin, which increases its crystalline phase. A direct heating process is a selection method that has been used by other researchers to produce geopolymer ceramics. On the contrary, the use of this method leads to production of geopolymer ceramics with lower mechanical properties; hence, the sintering method and temperature are the main factors that play a substantial role during transition phases in the geopolymer system.

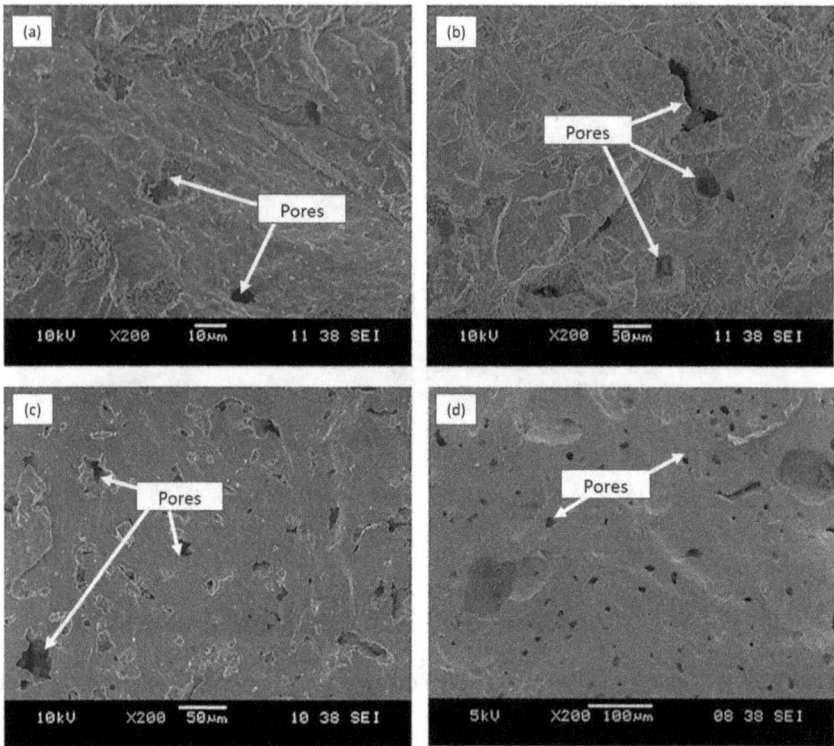

FIGURE 5.6   SEM micrograph of KGC and KGCB at different sintering temperatures: (a) 900°C, (b) 1,000°C, (c) 1,100°C, and (d) 1,200°C.

REFERENCES

Ahmad, R., Mustafa Al Bakri Abdullah, M., Mastura Wan Ibrahim, W., Mahyiddin Ramli, M., Victor Sandhu, A., Aida Mohd Mortar, N., & Wazien Ahmad Zailani, W. (2020). Geopolymer ceramic as piezoelectric materials: A review. *IOP Conference Series: Materials Science and Engineering, 864*(1), 012044.

Baklouti, S., Bouaziz, J., Chartier, T., & Baumard, J.-F. (2001). Binder burnout and evolution of the mechanical strength of dry-pressed ceramics containing poly(vinyl alcohol). *Journal of the European Ceramic Society, 21*(8), 1087–1092.

Barbosa, V. F. F., & MacKenzie, K. J. D. (2003). Thermal behaviour of inorganic geopolymers and composites derived from sodium polysialate. *Materials Research Bulletin, 38*(2), 319–331.

Bell, J. L., Driemeyer, P. E., & Kriven, W. M. (2009). Formation of ceramics from metakaolin-based geopolymers. Part II: K-based geopolymer. *Journal of the American Ceramic Society, 92*(3), 607–615.

Bernard-Granger, G., Monchalin, N., & Guizard, C. (2007). Sintering of ceramic powders: Determination of the densification and grain growth mechanisms from the "grain size/relative density" trajectory. *Scripta Materialia, 57*(2), 137–140.

Cong, P., & Cheng, Y. (2021). Advances in geopolymer materials: A comprehensive review. *Journal of Traffic and Transportation Engineering* (English Edition), 8(3), 283–314.

Davidovits, J., & Quentin, S. (1991). Geopolymers Inorganic polymerie new materials. *Journal of Thermal Analysis, 37,* 1633–1656.

Duxson, P., Provis, J. L., Lukey, G. C., Mallicoat, S. W., Kriven, W. M., & Van Deventer, J. S. J. (2005). Understanding the relationship between geopolymer composition, microstructure and mechanical properties. *Colloids and Surfaces A: Physicochemical and Engineering Aspects, 269,* 47–58.

Grigoriev, S. N., Fedorov, S. V., & Hamdy, K. (2019). Materials, properties, manufacturing methods and cutting performance of innovative ceramic cutting tools: A review. *Manufacturing Review, 6,* 19.

He, P., Yang, Z., Yang, J., Duan, X., Jia, D., Wang, S., Zhou, Y., Wang, Y., & Zhang, P. (2015). Preparation of fully stabilized cubic-leucite composite through heat-treating Cs-substituted K-geopolymer composite at high temperatures. *Composites Science and Technology, 107,* 44–53.

Heah, C. Y., & Hazlinda, K. (2011). Potential application of kaolin without calcine as greener concrete: A review. *Australian Journal of Basic and Applied Sciences, 5*(7), 1026–1035.

Hu, P., Gui, K., Hong, W., & Zhang, X. (2017). Preparation of ZrB_2-SiC ceramics by single-step and optimized two-step hot pressing using nanosized ZrB_2 powders. *Materials Letters, 200,* 14–17.

Iwahiro, T., Nakamura, Y., Komatsu, R., & Ikeda, K. (2001). Crystallization behavior and characteristics of mullites formed from alumina-silica gels prepared by the geopolymer technique in acidic conditions. *Journal of the European Ceramic Society, 21*(14), 2515–2519.

Iyasara, A. C., Joseph, M., Azubuike, T. C., & Tse, D. T. (2014). Exploring ceramic raw materials in nigeria and their contribution to nation's development. *American Journal of Engineering Research, 3*(9), 2320–2847.

Kovářík, T., Hájek, J., Pola, M., Rieger, D., Svoboda, M., Beneš, J., Šutta, P., Deshmukh, K., & Jandová, V. (2021). Cellular ceramic foam derived from potassium-based geopolymer composite: Thermal, mechanical and structural properties-NC-ND license. *Materials and Design, 198,* 109355. https:// creativecommons.org/licenses/by-nc-nd/4.0/.

Kriven, W. M., Zhou, Y., Radovic, M., Mathur, S., Ohji, T., & American Ceramic Society. (2010). Strategic materials and computational design. In: *Collection of Papers* Presented at *the 34th International Conference on Advanced Ceramics and Composites*, January 24–29, 2010, Daytona Beach, FL. John Wiley & Sons.

Mukherjee, S. (2013). Traditional and modern uses of ceramics, glass and refractories. In: *The Science of Clays* (pp. 123–150). Springer.

Olevsky, E., & Molinari, A. (2000). Instability of sintering of porous bodies. *International Journal of Plasticity, 16,* 1±37.

Palomo, A., Grutzeck, M. W., & Blanco, M. T. (1999). Alkali-activated fly ashes: A cement for the future. *Cement and Concrete Research, 29*(8), 1323–1329.

Peigang, H., Dechang, J., Meirong, W., & Yu, Z. (2011). Thermal evolution and crystallization kinetics of potassium-based geopolymer. *Ceramics International, 37,* 59–63.

Peigang, H., Dechang, J., & Shengjin, W. (2013). Microstructure and integrity of leucite ceramic derived from potassium-based geopolymer precursor. *Journal of the European Ceramic Society, 33*(4), 689–698.

Peigang, H., Dechang, J., Tiesong, L., Meirong, W., & Yu, Z. (2010). Effects of high-temperature heat treatment on the mechanical properties of unidirectional carbon fiber reinforced geopolymer composites. *Ceramics International, 36*(4), 1447–1453.

Pelz, J. S., Ku, N., Meyers, M. A., & Vargas-Gonzalez, L. R. (2021). Additive manufacturing of structural ceramics: A historical perspective. *Journal of Materials Research and Technology, 15,* 670–695.

Popoola, A., Olorunniwo, O., & Ige, O. (2014). Corrosion resistance through the application of anti-corrosion coatings. In: M. Aliofkhazraei (Ed.), *Developments in Corrosion Protection*. InTech.

Prado, M. O., Fredericci, C., & Zanotto, E. D. (2003). Non-isothermal sintering with concurrent crystallization of polydispersed soda-lime-silica glass beads. *Journal of Non-Crystalline Solids*, *331*(1–3), 157–167.

Rajeswari, K., Chaitanya, S., Biswas, P., Suresh, M. B., Rao, Y. S., & Johnson, R. (2015). Processing Research Binder burnout and sintering kinetic study of alumina ceramics shaped using methylcellulose. *Journal of Ceramic Processing Research*, *16*(1), 24–31.

Sames, W. J., List, F. A., Pannala, S., Dehoff, R. R., & Babu, S. S. (2016). The Metallurgy and processing science of metal. *International Materials Reviews*, *61*(5), 315–360.

Shahari, S., Fathullah, M., Abdullah, M. M. A. B., Shayfull, Z., Mia, M., Budi Darmawan, & V. E. (2021). Recent developments in fire retardant glass fibre reinforced epoxy composite and geopolymer as a potential fire-retardant material: A review. *Construction and Building Materials*, *277*, 12246.

Taktak, R., Baklouti, S., & Bouaziz, J. (2011). Effect of binders on microstructural and mechanical properties of sintered alumina. *Materials Characterization*, *62*(9), 912–916.

Wan Ibrahim, W. M., Bakri Abdullah, M. M. A., Ahmad, R., Mohd Tahir, M. F., Mohd Mortar, N. A., & Noor Azli, M. A. A. (2020). Mechanical and physical properties of bottom ash/fly ash geopolymer for pavement brick application. *IOP Conference Series: Materials Science and Engineering*, *743*(1), 012029.

Xie, N., Bell, J. L., & Kriven, W. M. (2010). Fabrication of structural leucite glass-ceramics from potassium-based geopolymer precursors. *The American Ceramic Society*, *93*(9), 2644–2649.

Xu, H., & Van Deventer, J. S. J. (2000). The geopolymerisation of alumino-silicate minerals. *International Journal of Mineral Processing*, *59*(3), 247–266.

Zawrah, M. F., Sadek, H. E. H., Ngida, E. A., Sawan, S. E. A., & El-Kheshen, A. A. (2022). Effect of low-rate firing on physico-mechanical properties of unfoamed and foamed geopolymers prepared from waste clays. *Ceramics International*, *48*, 11330–11337.

Zulkifli, N. I., Bakri Abdullah, M. M. A., Mohd Salleh, M. A. A., Ahmad, R., Sandu, A. V., & Mohd Mortar, N. A. (2020). Development of geopolymer ceramic as a potential reinforcing material in solder alloy: Short review. *IOP Conference Series: Materials Science and Engineering*, *743*(1), 012023.

6 Thermal Properties of Geopolymers

Liew Yun-Ming, Heah Cheng-Yong,
Mohd Mustafa Al Bakri Abdullah,
Ooi Wan-En, Ong Shee-Ween, Lim Jia-Ni,
Tee Hoe-Woon, and Nur Ain Jaya
Universiti Malaysia Perlis

Wei-Hao Lee
National Taipei University of Technology

6.1 INTRODUCTION

Geopolymers, as inorganic materials, are also called alkali-activated materials. It has an amorphous structure consisting of a three-dimensional polymeric chain of Si–O–Al bonds. Geopolymer can be prepared by alkali activation of aluminosilicate materials at room temperature or higher temperature. These materials mostly comprise amorphous silica SiO_2 and Al_2O_3 with high pozzolanic activity. Apart from natural sources (kaolin and metakaolin), industrial by-products such as fly ash, bottom ash, and blast furnace slag can be used as aluminosilicate materials. The ceramic-like properties of geopolymers have been widely reported to render good thermal properties which make geopolymers suitable to be used in thermal insulation, automotive, and refractory applications. The thermal properties of geopolymers include:

a. Physical changes (explosive spalling, pore size, and density)
b. Residual strength or strength retention
c. Thermal expansion and shrinkage
d. Thermal conductivity
e. Microstructural changes (degree of crystallinity, phase, distribution of pore size, and weight change)

In this chapter, the content focuses on the thermal expansion and shrinkage, thermal conductivity, weight change, and heat evolution of geopolymers. The physical and mechanical properties changes and other thermochemical-related properties will be discussed. The thermal properties of geopolymers could be accessed by differential scanning calorimetry (DSC), differential temperature analysis (DTA), and thermogravimetry analysis (TGA). DSC and DTA techniques are used to analyze the transformation of geopolymers at high temperatures. The TGA technique measures the mass loss when the sample is gradually heated up to elevated temperatures.

DOI: 10.1201/9781003390190-6

The key determinants of the thermal properties of geopolymers are water content and pore distribution, as well as the amorphous and crystalline phases in the geopolymer structure and their nominal composition. The presence of humidity and interconnected pore structures reflects the thermal transport properties of geopolymers.

When heat is applied, water is expelled from geopolymers. Generally, major mass loss observed in the TGA curve usually occurs at 100°C–300°C due to the evaporation of free water. Further weight loss at higher temperatures (300°C–700°C) is attributed to the evaporation of chemically bound water and the structural hydroxyl groups. As illustrated in the TGA curve (Figure 6.1), the mass loss of geopolymer can mainly be separated into four stages, wherein most of the mass loss occurs at a temperature below 265°C. Stage 1 took place at 25°C–110°C, resulting from the evaporation of free water, while stage 2 (110°C–265°C) corresponds to the evaporation of physically bound water. The mass loss at stage 3 (270°C–630°C) is due to the evaporation of physico-chemically bound water and the dihydroxylation polycondensation process of geopolymeric gel. Stage 4 (680°C–1,000°C) is attributed to the calcination decomposition of inorganic carbonate salt, muscovite, and kaolinite during sintering instead of further dihydroxylation of the gel reaction product (Liu et al., 2020).

The thermal performance of pure geopolymer and nano Al_2O_3 geopolymer specimens was compared by Xavier and Rahim (2022), as shown in Figure 6.2. A significant weight loss was found at ambient temperature and 150°C for pure geopolymer and nano Al_2O_3 geopolymer, owing to the evaporation of physiologically adsorbed water. The DTG curve of pure geopolymer is steeper than that of nano Al_2O_3 geopolymer. This is because the pure geopolymer maximum migrated to a lower temperature than the nano Al_2O_3 geopolymer point.

A significant finding was found by Nath and Kumar (2020), in which the mass loss that resulted from the de-carbonation of secondary carbonate compounds and dehydroxylation of crystalline zeolitic phases could be reduced with the milling time

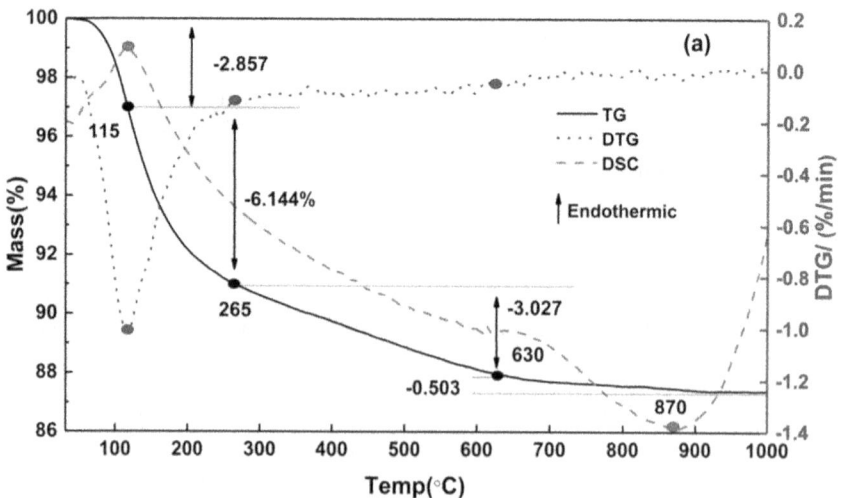

FIGURE 6.1   Mass loss and heat flow rate curves of metakaolin geopolymers (Liu et al., 2020).

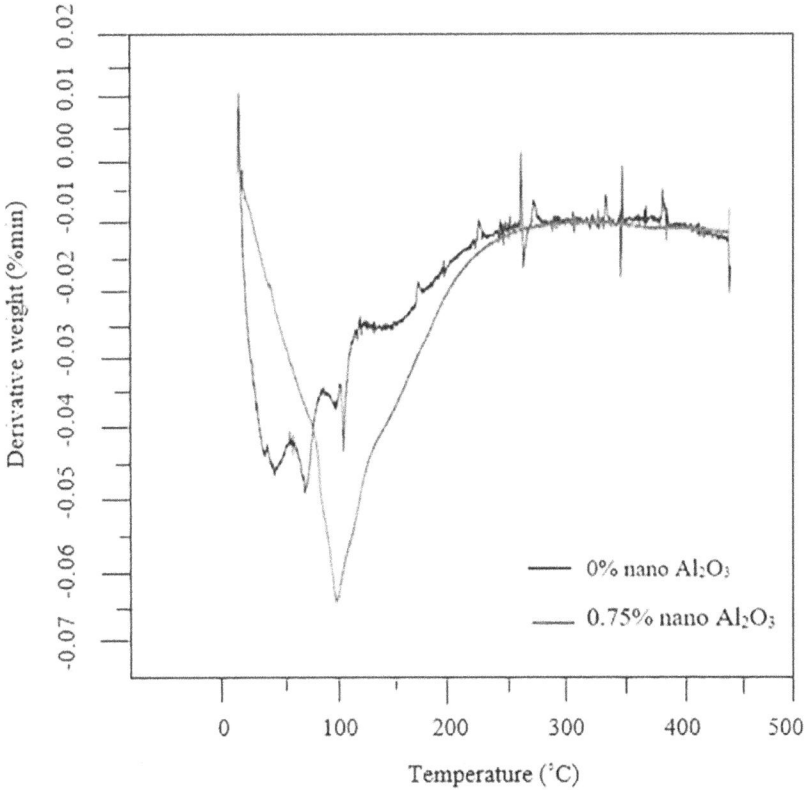

FIGURE 6.2 DTG curves of metakaolin and fly ash geopolymers (Xavier and Rahim, 2022).

of fly ash. This is reasonable since the increased milling time enhanced the degree of fineness of particles, which subsequently increased the reactivity of raw materials. As the finer fraction is more reactive, most of the alkali activator is consumed in the geopolymerization reaction, and a lesser amount is left in the composite body to react with carbon dioxide. Hence, the carbonation is lowered since carbon dioxide does not penetrate the resulting geopolymer matrix.

Furthermore, past investigations confirmed that the addition of materials could affect the thermal stability of geopolymers. Likewise, Liang et al. (2019) reported that the geopolymer with the inclusion of rice husk ash exhibited a sharp reduction in mass loss in comparison to pure geopolymer. Rice husk ash microparticles play an important role in filling the micro-cracks and micropores of geopolymer inner space, yielding a denser microstructure. The dense microstructure reduces the free and bound water that existed throughout gel phases and, thus, lowers the mass loss of geopolymer.

Meanwhile, Tan et al. (2022) have tested the influence of municipal solid waste incineration fly ash addition on the reaction heat evaluation of construction and demolition waste-based geopolymers. The TGA results (Figure 6.3) proved that the addition of fly ash reduced the mass loss that occurred between 30°C and 300°C. The inclusion of fly ash provokes the formation of C-A-S-H gels, which

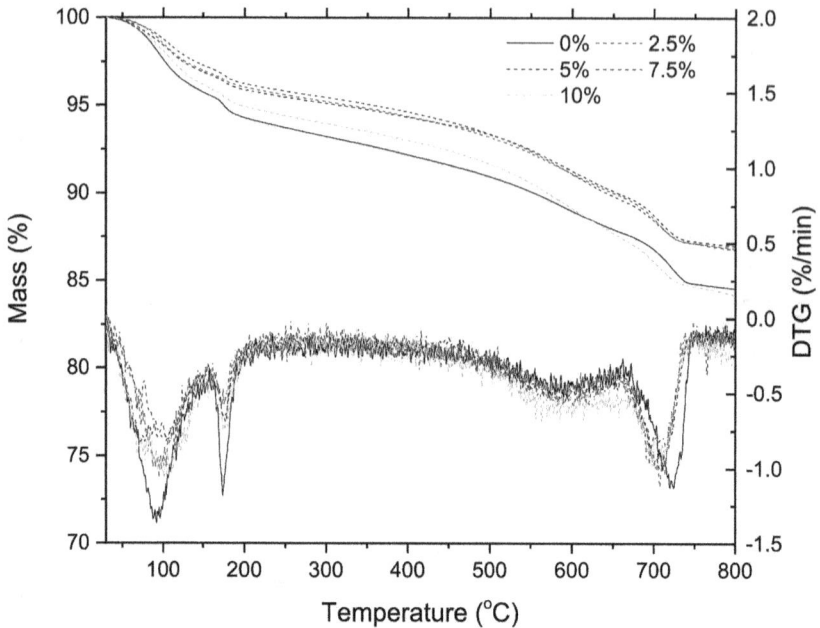

FIGURE 6.3 TGA/DTG thermograms of construction and demolition waste geopolymers (Tan et al., 2022).

fill in the microvoids and enhance the density of the gel framework. Not only that, the lower degree of gel development in fly ash-containing samples reduced the availability of unbound water in the system. Therefore, it can be said that the inclusion of fly ash promotes the formation of gel in construction and demolition waste geopolymers.

The decrease or increase in mass loss in the TGA curve is represented as endothermic or exothermic peaks under the DSC and DTA curves. The water content and its chemical form are represented by the energy absorption under the DSC thermogram. The presence of water in geopolymer structures is closely related to their thermal properties. Kaze et al. (2021) prepared iron-rich laterite geopolymers with a sodium-based alkali activator. The DTA curve of geopolymer in Figure 6.4 shows less pronounced endothermic peaks at 100°C that are associated with the escape of free or physiosorbed water. A noticeable endothermic peak observed at 150°C is due to the gradual destruction of geopolymer structure as a result of structural water evaporation. At 600°C, the kaolinite mineral in iron-rich laterites is completely transformed into metakaolinite.

Based on Yang et al. (2017), the thermal stability of geopolymers is dependent on the chemical structure of the hardened geopolymers and the absence of carbonates and chemically bound water-containing compounds. Minimal water evaporation prevents cracking and thus contributes to better structural stability (Liu et al., 2020).

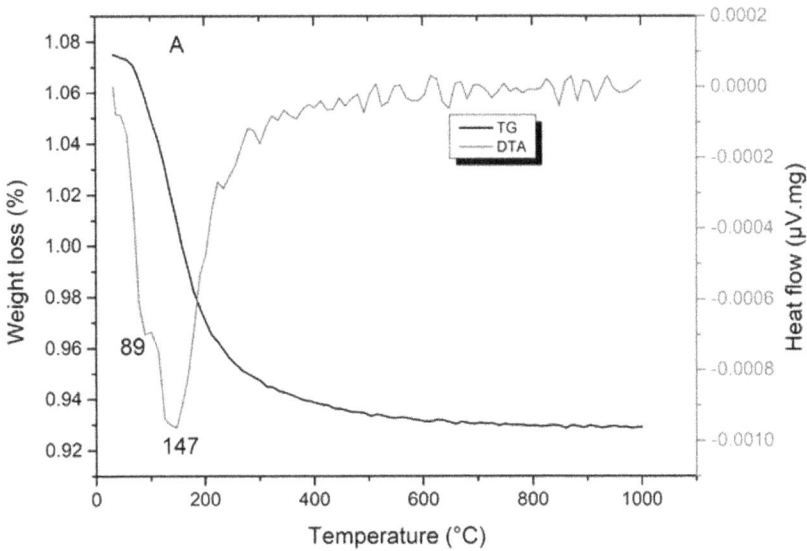

FIGURE 6.4 DTA curve of iron-rich laterites geopolymer (Kaze et al., 2021).

6.2 THERMAL EXPANSION AND SHRINKAGE ANALYSIS

Thermal expansion and shrinkage induce both external and internal stress on heated samples, which would cause destruction and damage to the geopolymer structure. Usually, thermal expansion or shrinkage occurs due to the non-uniformity of the composition when subjected to a temperature change. The samples would experience physical-chemical evolution during the thermal deformation analysis executed in a dilatometer (usually from 20°C to 1,300°C).

As referred to in Figure 6.5, the dilatometric curve can normally be divided into four regions, indicating the four stages of thermal shrinking of the geopolymer, namely (A) structural resilience, (B) dehydration, (C) dehydroxylation, and (D) sintering. The shrinkage in region A (from room temperature to 100°C) is due to free water evaporation. Shrinkage occurs in region B (100°C–300°C) due to the loss of interstitial water from nanopores and micropores, while shrinkage in region C (300°C–800°C) is attributed to the condensation reaction. The sintering process happens in region D (>800°C), where the amorphous phase of the geopolymer melts into the glass phase, leading to further densification. The shrinkage is severe if the initial water content is higher (Chen et al., 2022; Kuenzel et al., 2012).

Numerous studies have implemented various fabrication methods or incorporation of various materials to improve the thermal stability of the geopolymer. For instance, mullite hindered the shrinkage of geopolymers during heat treatment. According to Boum et al. (2020), mixing 30% of alumina-rich bauxite in metakaolin-based geopolymer favors the formation of liquid-phase mullite crystals in the matrix, subsequently reducing the extent of shrinkage at 1,200°C. Likewise, the incorporation

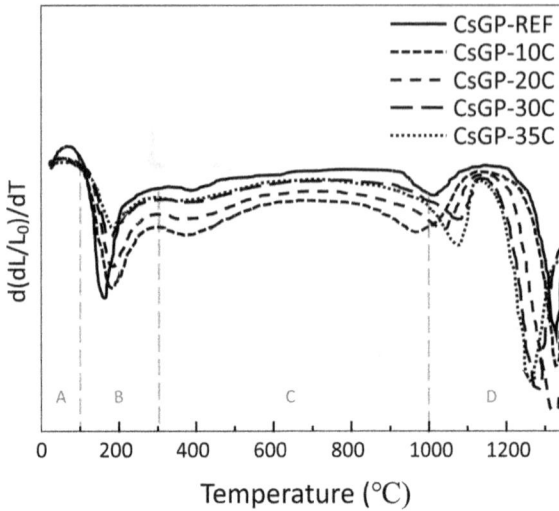

FIGURE 6.5 Dilatometric curve of cesium-activated geopolymer composited (CsGPc) with 5–35 wt.% cordierite reinforcements (Chen et al., 2022).

of 20% mullite fiber in metakaolin geopolymer reduced the thermal shrinkage by 30% at 1,450°C, as presented by Wei et al. (2022). These are due to the presence of thermally resistant crystals that contribute to the volumetric thermal stability of geopolymers (Klima et al., 2022).

Furthermore, cordierite powder with low thermal expansion and outstanding thermal shock resistance reduced the thermal shrinkage of metakaolin geopolymer activated by cesium silicate. A remarkable finding was revealed by Chen et al. (2022), where a reduction of shrinkage up to 70% was achieved as cordierite contained the volatile cesium above 1,100°C. Meanwhile, the silane coupling agent, a colorless liquid, reduces the shrinkage rate of metakaolin-based geopolymer below 250°C (Wang et al., 2021). It is found that the hydration of the silane coupling agent reacted with Si–OH and Al–OH bonds from the metakaolin, which extended the formation of geopolymer gel. The scenario reduces the free water present in the geopolymer matrix and lowers the evaporation of free water when the temperature is raised. Therefore, shrinkage is not evident. When the temperature rises beyond 250°C, the condensation reaction of geopolymer occurs, and the silanol that is bound to the geopolymer disappears, resulting in greater shrinkage of the metakaolin geopolymer.

Aggregates influence the thermal stability of geopolymer mortar/concrete. At high temperatures, geopolymer paste undergoes sintering, densification, and strength gain, but aggregates may experience expansion/shrinkage simultaneously. The compatibility of geopolymer paste and aggregates strongly influences their performance at elevated temperatures. An opposing behavior may result in crack formation in the interfacial transition zone, negatively affecting the physical and mechanical properties of the geopolymer mortar/concrete. Abdulkareem et al. (2014) tested the thermal expansion of fly ash geopolymers and lightweight expanded clay aggregate in order

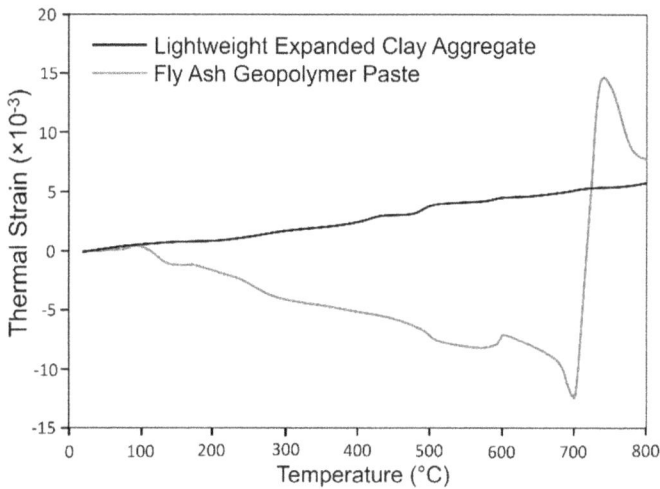

FIGURE 6.6   Dilatometer curve of fly ash geopolymer and lightweight expanded clay aggregate (Abdulkareem et al., 2014).

to access the thermal mismatch between them in geopolymer concrete, as shown in Figure 6.6. The geopolymer paste shows little expansion at 70°C–100°C, followed by sharp shrinkage up to 700°C and sharp expansion at 700°C–800°C. Based on Rickard et al. (2013), the expansion at 700°C–800°C is usually accompanied by a drastic shrinkage as the result of viscous flow and sintering of the aluminosilicate materials, which cause densification of geopolymers. On the other hand, the lightweight aggregate expands minimally up to 800°C and thus has a low coefficient of thermal expansion (CTE). The thermal mismatch between geopolymer paste and the lightweight aggregate contributes to micro-cracks in the interfacial transition zone. Even so, as the concrete comprises 75% of the lightweight aggregate, the strength degradation at elevated temperatures is low (Abdulkareem et al., 2014).

Based on Sivasakthi et al. (2021), the expansion of river sand is five times higher than that of the copper slag at 1,000°C. The geopolymer mortar containing river sand experienced the greatest expansion at 700°C, with about 1.7% linear thermal expansion. Meanwhile, the geopolymer mortar containing copper slag undergoes negligible thermal shrinkage up to only 0.008% throughout the temperature range of 30°C–1,000°C. The superior thermal stability is attributed to the slow dehydroxylation of geopolymer gel when copper slag is applied. Subsequently, the compressive strength of the geopolymer mortar containing copper slag is generally higher than that of river sand after exposure to elevated temperatures.

Similarly, wollastonite enhanced the thermal stability of fly ash-based geopolymers (Hemra et al., 2021), as shown in Figure 6.7. Significant thermal shrinkage occurred in the geopolymer paste. In contrast, the geopolymer is composited with 30 and 50 wt.% wollastonite and expands slightly from 25°C up to 900°C, enhancing the thermal stability of the composites.

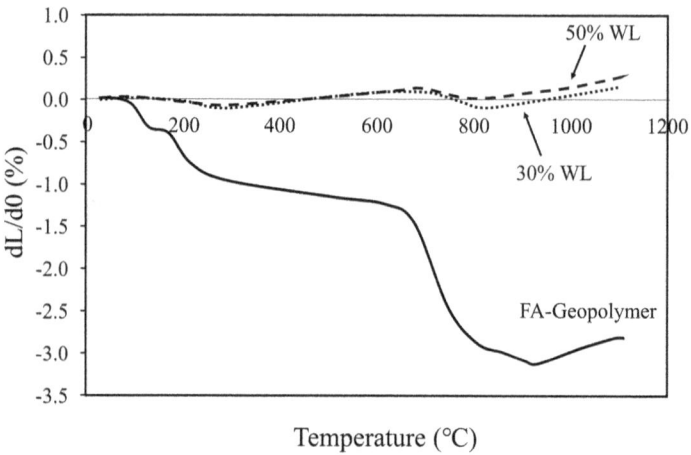

FIGURE 6.7 Thermal expansion of fly ash geopolymer paste and composites (Hemra et al., 2021).

6.3 THERMAL CONDUCTIVITY ANALYSIS

The thermal conductivity (k or λ) of a material is a measurement of its effectiveness to conduct heat. It is defined as the amount of heat transmitted per unit thickness and unit area per unit temperature gradient. A lower k value represents lower heat conduction, while a higher k value denotes a lower temperature gradient and thus lower thermal stress. The thermal conductivity can be modified by incorporating varying materials or implying different production approaches in the development of geopolymers.

Generally, geopolymer is a poorer heat-conducting material compared to conventional Portland cement due to its higher porosity and lower chemically bound water (Jaya et al., 2020), where the thermal conductivity of air is 25 times lower than that of water (Sivasakthi et al., 2021). Geopolymer usually has half the thermal conductivity (λ) value of conventional Portland cement (1.5 W/m K). To enhance the heat transfer of geopolymer, the geopolymer could be made denser or porous (increase or reduce density) by modifying the mix proportion. As demonstrated by Jaya et al. (2020), increasing the Na_2SiO_3/NaOH of metakaolin-based geopolymer from 0.4 to 1.0 reduces the porosity by approximately 6% and leads to a rise in thermal conductivity of about 8 W/m K. Porosity is the most influential parameter of thermal conductivity. With reduced tortuosity and increased continuous gel structure, more heat transfer is permitted, and thus thermal conductivity can be increased. Likewise, increasing the SiO_2/Al_2O_3 ratio from 1.3 to 2.5 increased the k values from 0.28 to 0.35 W/m K for metakaolin geopolymers due to the reduction in pore size (Kamseu et al., 2012).

Another way to increase the thermal conductivity of geopolymers is by adding thermally conductive elements. But phonon scattering occurs at the interface of geopolymer composite, causing resistance in heat transfer. Therefore, it is also important to build an effective phonon transport pathway to optimize the effect of the thermally conductive fillers. The incorporation of carbon nanotubes in the geopolymer matrix performed by Zhu et al. (2021) successfully overcomes the issue. The thermal conductivity of the metakaolin-based geopolymer composited with 5 vol.% carbon nanotubes coated with silica is increased by 71.7% compared to the neat geopolymer. Silica coating is implemented to overcome the agglomeration issue as well as the phonon scattering concern. Such an approach establishes an effective heat conduction network in the geopolymer matrix.

Nonetheless, most of the research in geopolymer aims to widen the range of thermal-resistant applications, especially for exterior building elements. Several methods are implemented to reduce the thermal conductivity of geopolymers, such as the introduction of pores in the geopolymer structure and the addition of lightweight and/ or porous aggregates/fillers.

Geopolymer foams are highly porous geopolymer materials that can be developed using foaming admixtures. The air voids in foamed geopolymer act as a heat flow barrier across the specimen and can significantly improve the insulation properties. Hydrogen peroxide (H_2O_2) is one of the most popular foaming agents. Based on Pantongsuk et al. (2021), the increase of H_2O_2 from 0.5 to 1.5 wt.% in the metakaolin and bagasse ash-blended geopolymer foams yields a significant increase in the porosity, followed by a significant reduction in thermal conductivity, from approximately 0.40 to 0.16 W/m K. Similarly, adding 1.2 wt.% of H_2O_2 to biomass fly ash-based geopolymers yields a thermal conductivity of 0.107 W/m K (Novais et al., 2016). Rice husk ash that is rich in silicon dioxide (SiO_2) is also a forming agent for geopolymers. According to Liang et al. (2022), rice husk ash has lower thermal conductivity and can cultivate fine pores in the geopolymer matrix. The slag-based geopolymer foamed by rice husk ash reveals thermal conductivity ranging from 0.110 to 0.289 W/m K, which exhibits good thermal insulation characteristics.

Furtos et al. (2021) reduce the thermal conductivity of fly ash-based geopolymer from 0.43 to 0.22 W/m K by increasing the substitution of fly ash by wood fiber up to 35%. The insulating properties of light-weighted wood fiber reduce the density and thermal conductivity of wood fiber-reinforced geopolymer composites. Thus, the increase in wood fiber content reduces the thermal conductivity. However, the interaction between geopolymer and fibers with low melting points triggers distinctive performance. Based on Agustini et al. (2021), the effect of polypropylene (PP) fiber on fly ash-based geopolymer foams varies according to the thermal loading. The evolution of thermal conductivity of the geopolymer with 0%, 0.25%, and 0.50% of PP fiber by the mass of fly ash is illustrated in Figure 6.8. Generally, adding PP fiber to the geopolymer foams fills up the voids in the structure and increases the density, subsequently raising the thermal conductivity. When subjected to thermal loading beyond 100°C, the PP fiber melts and incurs extra voids in the matrix of geopolymer foams, increasing the resistance to heat transfer and further reducing the thermal conductivity of the foams.

Besides fibers, the incorporation of light-weighted aggregates has a significant influence on the thermal properties of the mortar/concrete. Normal-weight concrete has a thermal conductivity of 1.98 W/m K (Real et al., 2016). The introduction of light-weighted aggregates with a density of less than 2.0 kg/m³ can improve the thermal resistivity of mortar/concrete. Geopolymer can be treated as a lightweight aggregate in the production of mortar/cement. Geopolymer aggregate was produced using

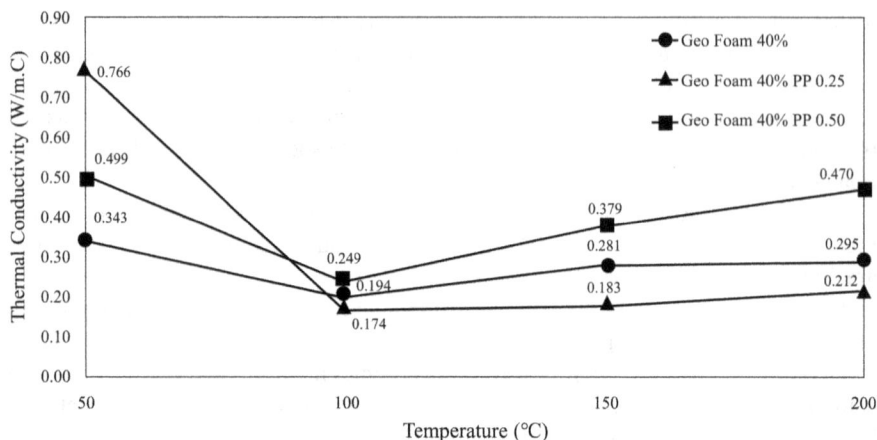

FIGURE 6.8 Thermal conductivity of geopolymer foams with a varied amount of PP fiber from 50°C to 200°C (Agustini et al., 2021).

palm oil fuel ash and sludge that have a bulk density of 1.12–1.18 kg/m³ (Kwek et al., 2022). The application of the geopolymer aggregate successfully reduced the thermal conductivity of concrete to 1.04 W/m K. Moreover, expanded perlite is a porous aggregate that can be an alternate aggregate for sand. The incorporation of expanded perlite improves thermal resistance by delivering micropores in geopolymer concrete. The increase of expanded perlite from 0% to 20% reduces the thermal conductivity of geopolymer foamed concrete by 12.1% (Pasupathy et al., 2020). Other potential fillers/aggregates are listed in Table 6.1 along with their respective thermal conductivity, λ, of the composited geopolymer.

Overall, the thermal conductivity of geopolymers is significantly influenced by the density, porosity, and pore size of the materials. An increase in the porosity of the geopolymers leads to a decrease in their thermal conductivity. However, this is in regard to the fact that the porosity is homogenously distributed and does not form large pores that form thermal bridges, which increase the thermal conductivity. The thermal conductivity of geopolymers can be carefully tailored to suit their applications. In addition, the incorporation of lightweight aggregates, foaming agents, or porous aggregates could reduce the thermal conductivity values.

6.4 CONCLUSIONS

Extensive research has been conducted on the thermal characteristics of geopolymers. The effects of temperature variations on geopolymers include alterations in mass and changes in size, either through expansion or shrinkage. These responses are attributed to the type of geopolymer matrix and the thermal stability of the raw materials used. A denser geopolymer matrix with less bound and unbound water can decrease mass loss and cracking, and this can be achieved by modifying the water content, types of aluminosilicate materials, alkali activators, admixtures, and additives used. In addition, the thermal stability of fillers and aggregates significantly affects the thermal expansion and shrinkage of geopolymer composites. Using raw materials with higher thermal stability or lower thermal mismatch with

TABLE 6.1
Thermal Conductivity (λ) of Geopolymer Composited with Various Fillers/Aggregates (Zhang et al., 2021)

Fillers/Aggregates	λ (W/m.K)
Fly ash cenospheres	0.17–0.36
Waste glass fibers	0.22–0.69
Wheat straw	0.09–0.19
Monofilament polypropylene fibers	0.26–0.35
Sawdust	0.15–0.35
Expanded vermiculite	0.18–0.26
Macro-encapsulated aggregates	0.21
Epoxy resin	0.10–0.11
Oligomeric dimethylsiloxane mixture	0.10–0.11
Polystyrene particles	0.03–0.08
Silica sand and polyvinyl alcohol (PVA) fibers	1.85
Expanded glass and PVA fibers	0.93
Ceramic microspheres and PVA fibers	1.14
Expanded perlite and PVA fibers	1.10
SiC powder	0.15
Glass microspheres	0.05–0.08
Waste polyurethane powders	0.30–0.36
Waste expanded polystyrene, marble powder, and epoxy resin	0.12–0.21
Silica aerogel	0.05–0.07
Basalt fiber	0.13–0.36
Polypropylene fiber and glazed hollow beads	0.15–0.58
Cenospheres	0.29
Sawdust	0.09–0.27
Zeolite, bentonite	0.12–0.17
Marble waste, acrylic fibers, and cotton fibers	0.34–0.91

the geopolymer matrix can reduce physical damage. To expand the use of geopolymer for thermal insulation purposes, insulation materials were incorporated, or foam was induced into the geopolymer structure to reduce the thermal conductivity of the material. In summary, extensive research on the thermal properties of geopolymers has led to the development of materials with improved properties and expanded application possibilities.

REFERENCES

Abdulkareem, O. A., Mustafa Al Bakri, A. M., Kamarudin, H., Khairul Nizar, I., & Saif, A. A. (2014). Effects of elevated temperatures on the thermal behavior and mechanical performance of fly ash geopolymer paste, mortar and lightweight concrete. *Construction and Building Materials*, *50*, 377–387. doi:10.1016/j.conbuildmat.2013.09.047

Agustini, N. K., Triwiyono, A., Sulistyo, D., & Suyitno, S. (2021). Mechanical properties and thermal conductivity of fly ash-based geopolymer foams with polypropylene fibers. *Applied Sciences*, *11*(11). doi:10.3390/app11114886

Boum, R. B. E., Kaze, C. R., Nemaleu, J. G. D., Djaoyang, V. B., Rachel, N. Y., Ninla, P. L., & Kamseu, E. (2020). Thermal behaviour of metakaolin-bauxite blends geopolymer: Microstructure and mechanical properties. *SN Applied Sciences, 2*(8), 1358. doi:10.1007/s42452-020-3138-9

Chen, W., Garofalo, A. C., Geng, H., Liu, Y., Wang, D., & Li, Q. (2022). Effect of high temperature heating on the microstructure and performance of cesium-based geopolymer reinforced by cordierite. *Cement and Concrete Composites, 129*, 104474. doi:10.1016/j.cemconcomp.2022.104474

Furtos, G., Molnar, L., Silaghi-Dumitrescu, L., Pascuta, P., & Korniejenko, K. (2021). Mechanical and thermal properties of wood fiber reinforced geopolymer composites. *Journal of Natural Fibers*, 1–16. doi:10.1080/15440478.2021.1929655

Hemra, K., Kobayashi, T., Aungkavattana, P., & Jiemsirilers, S. (2021). Enhanced mechanical and thermal properties of fly ash-based geopolymer composites by wollastonite reinforcement. *Journal of Metals, Materials and Minerals, 31*(4), 13–25.

Jaya, N. A., Yun-Ming, L., Cheng-Yong, H., Abdullah, M. M. A. B., & Hussin, K. (2020). Correlation between pore structure, compressive strength and thermal conductivity of porous metakaolin geopolymer. *Construction and Building Materials, 247*, 118641. doi:10.1016/j.conbuildmat.2020.118641

Kamseu, E., Ceron, B., Tobias, H., Leonelli, E., Bignozzi, M. C., Muscio, A., & Libbra, A. (2012). Insulating behavior of metakaolin-based geopolymer materials assess with heat flux meter and laser flash techniques. *Journal of Thermal Analysis and Calorimetry, 108*(3), 1189–1199. doi:10.1007/s10973-011-1798-9

Klima, K. M., Schollbach, K., Brouwers, H. J. H., & Yu, Q. (2022). Enhancing the thermal performance of Class F fly ash-based geopolymer by sodalite. *Construction and Building Materials, 314*, 125574. doi:10.1016/j.conbuildmat.2021.125574

Kuenzel, C., Vandeperre, L. J., Donatello, S., Boccaccini, A. R., & Cheeseman, C. (2012). Ambient temperature drying shrinkage and cracking in metakaolin-based geopolymers. *Journal of the American Ceramic Society, 95*(10), 3270–3277.

Kwek, S. Y., Awang, H., Cheah, C. B., & Mohamad, H. (2022). Development of sintered aggregate derived from POFA and silt for lightweight concrete. *Journal of Building Engineering, 49*, 104039. doi:10.1016/j.jobe.2022.104039

Liang, G., Liu, T., Li, H., Dong, B., & Shi, T. (2022). A novel synthesis of lightweight and high-strength green geopolymer foamed material by rice husk ash and ground-granulated blast-furnace slag. *Resources, Conservation and Recycling, 176*, 105922. doi:10.1016/j.resconrec.2021.105922

Liang, G., Zhu, H., Zhang, Z., & Wu, Q. (2019). Effect of rice husk ash addition on the compressive strength and thermal stability of metakaolin based geopolymer. *Construction and Building Materials, 222*, 872–881.

Liu, X., Jiang, J., Zhang, H., Li, M., Wu, Y., Guo, L., Wang, W., Duan, P., & Zhang, Z. (2020). Thermal stability and microstructure of metakaolin-based geopolymer blended with rice husk ash. *Applied Clay Science, 196*, 105769. doi:10.1016/j.clay.2020.105769

Nath, S. K., & Kumar, S. (2020). Role of particle fineness on engineering properties and microstructure of fly ash derived geopolymer. *Construction and Building Materials, 233*, 117294. doi:10.1016/j.conbuildmat.2019.117294

Novais, R. M., Buruberri, L. H., Ascensão, G., Seabra, M. P., & Labrincha, J. A. (2016). Porous biomass fly ash-based geopolymers with tailored thermal conductivity. *Journal of Cleaner Production, 119*, 99–107. doi:10.1016/j.jclepro.2016.01.083

Pantongsuk, T., Kittisayarm, P., Muenglue, N., Benjawan, S., Thavorniti, P., Tippayasam, C., Nilpairach, S., Heness, G., & Chaysuwan, D. (2021). Effect of hydrogen peroxide and bagasse ash additions on thermal conductivity and thermal resistance of geopolymer foams. *Materials Today Communications, 26*, 102149. doi:10.1016/j.mtcomm.2021.102149

Pasupathy, K., Ramakrishnan, S., & Sanjayan, J. (2020). Enhancing the mechanical and thermal properties of aerated geopolymer concrete using porous lightweight aggregates. *Construction and Building Materials*, *264*, 120713. doi:10.1016/j.conbuildmat.2020.120713

Real, S., Gomes, M. G., Moret Rodrigues, A., & Bogas, J. A. (2016). Contribution of structural lightweight aggregate concrete to the reduction of thermal bridging effect in buildings. *Construction and Building Materials*, *121*, 460–470. doi:10.1016/j.conbuildmat.2016.06.018

Rickard, W. D. A., Vickers, L., & van Riessen, A. (2013). Performance of fibre reinforced, low density metakaolin geopolymers under simulated fire conditions. *Applied Clay Science*, *73*, 71–77. doi:10.1016/j.clay.2012.10.006

Rodrigue Kaze, C., Ninla Lemougna, P., Alomayri, T., Assaedi, H., Adesina, A., Kumar Das, S., Lecomte-Nana, G., Kamseu, E., Melo, U. C., & Leonelli, C. (2021). Characterization and performance evaluation of laterite based geopolymer binder cured at different temperatures. *Construction and Building Materials*, *270*, 121443. doi:10.1016/j.conbuildmat.2020.121443

Sivasakthi, M., Jeyalakshmi, R., & Rajamane, N. P. (2021). Fly ash geopolymer mortar: Impact of the substitution of river sand by copper slag as a fine aggregate on its thermal resistance properties. *Journal of Cleaner Production*, *279*, 123766. doi:10.1016/j.jclepro.2020.123766

Tan, J., De Vlieger, J., Desomer, P., Cai, J., & Li, J. (2022). Co-disposal of construction and demolition waste (CDW) and municipal solid waste incineration fly ash (MSWI FA) through geopolymer technology. *Journal of Cleaner Production*, *148*, 132502.

Wang, X., Zhang, C., Wu, Q., Zhu, H., & Liu, Y. (2021). Thermal properties of metakaolin-based geopolymer modified by the silane coupling agent. *Materials Chemistry and Physics*, *267*, 124655. doi:10.1016/j.matchemphys.2021.124655

Wei, Q., Liu, Y., & Le, H. (2022). Mechanical and thermal properties of phosphoric acid activated geopolymer materials reinforced with mullite fibers. *Materials*, *15*(12), 4185.

Xavier, C. S. B., & Rahim, A. (2022). Nano aluminium oxide geopolymer concrete: An experimental study. *Materials Today: Proceedings*, *56*, 1643–1647. doi:10.1016/j.matpr.2021.10.070

Yang, T., Wu, Q., Zhu, H., & Zhang, Z. (2017). Geopolymer with improved thermal stability by incorporating high-magnesium nickel slag. *Construction and Building Materials*, *155*, 475–484. doi:10.1016/j.conbuildmat.2017.08.081

Zhang, X., Bai, C., Qiao, Y., Wang, X., Jia, D., Li, H., & Colombo, P. (2021). Porous geopolymer composites: A review. *Composites Part A: Applied Science and Manufacturing*, *150*, 106629. doi:10.1016/j.compositesa.2021.106629

Zhu, Y., Qian, Y., Zhang, L., Bai, B., Wang, X., Li, J., Bi, S., Kong, L., Liu, W., & Zhang, L. (2021). Enhanced thermal conductivity of geopolymer nanocomposites by incorporating interface engineered carbon nanotubes. *Composites Communications*, *24*, 100691. doi:10.1016/j.coco.2021.100691

7 Elevated Temperatures Exposure of Geopolymers

Heah Cheng-Yong, Liew Yun-Ming,
Mohd Mustafa Al Bakri Abdullah,
Khairunisa Zulkifly, Ng Hui-Teng,
Hang Yong-Jie, and Ng Yong-Sing
Universiti Malaysia Perlis

Wei-Hao Lee
National Taipei University of Technology

7.1 INTRODUCTION

The current construction structure takes care of safety and strength in relation to room temperature levels. Exposure to elevated temperatures aids in assessing how a construction structure will respond to a combination of high temperatures, which imitates direct fire exposure in real life. In the last few decades, it has been undeniable that geopolymers have developed great mechanical strength with regard to high-temperature gradients (Davidovits, 2020). According to the literature, the reaction between aluminosilicate precursors, such as fly ash (FA), ground granulated blast furnace slag (GGBFS), ladle furnace slag (LFS), metakaolin (MK), and electric arc furnace slag (EAFS), and activator solution (mixture of alkali hydroxide and alkali silicate), has potential for fire protection applications due to its excellent strength retention and outstanding high-temperature endurance.

In general, the main concern of current ordinary Portland cement (OPC) is its high degree of fire resistance. Although OPC concrete is an incombustible material, during fire accidents it suffers serious damage, including concrete spalling and mechanical property degradation. Compared to OPC, numerous investigations have shown that the geopolymer's inorganic framework makes up its thermal resistance. Davidovits (2020) has found that after geopolymerizing their water to a temperature of 200°C, followed by further geopolymerization, they can still retain their X-ray amorphous tetrahedral alumina (Al) and silica (Si) network up to 1,300°C.

This chapter discusses the physical properties (physical appearance changes, mass loss, volume changes, and density changes), mechanical properties (compressive strength and flexural strength), and material characterization (microstructure changes and phase changes) of the geopolymers upon elevated temperature exposure. The key determinants of the effect of elevated temperature exposure on geopolymer are the moisture content and porosity within the geopolymer matrix. Besides the internal

DOI: 10.1201/9781003390190-7

factors, such as the proportions of the raw material and alkali activator, the molarity of the alkali solution, and the aluminosilicate-to-alkali ratio, the external factors, such as temperature and exposure time, would also affect the performance of the geopolymer.

7.2 PHYSICAL PROPERTIES

The physical defect of geopolymers upon thermal exposure to different temperatures is manifested by the change in the geopolymer's appearance, mass, volume, and density. The physical evolution of the geopolymer during thermal exposure plays an important role in evaluating the performance of the geopolymer at elevated temperatures.

7.2.1 PHYSICAL APPEARANCE ANALYSIS

The geopolymer's surface changes of FA and FA-GGBFS geopolymers upon thermal exposure are evaluated by Hager et al. (2021), as illustrated in Figure 7.1. Color change was observed during high-temperature exposure, relating to oxidation of Fe_2O_3 in the aluminosilicate precursor. At the same time, the surface of FA geopolymers showed more pores compared to the surface of FA-GGBFS geopolymers. The inclusion of GGBFS aided in reducing the porosity of the geopolymer. The growth of pores in both geopolymers was linked to the dehydration process during the heating process. Moreover, no sign of spalling or cracking was observed on the surface of the geopolymers.

Other geopolymers, such as GGBFS, MK, and GGBFS-MK geopolymers, and the OPC were produced by Burciaga-Díaz and Escalante-García (2017) in order to compare their physical appearance at 600°C, 1,000°C, and 1,200°C. The surface of the geopolymer showed some minor cracks and pores at 600°C and 1,000°C. At 1,200°C, the surface of OPC remained the same, but a denser surface was obtained on MK geopolymer, which was related to the densification effect during the heat treatment. GGBFS geopolymer had severe cracks on its surface at 1,200°C which was in line with the thermal expansion caused by the phase transformation. The GGBFS-MK geopolymer melted at 1,200°C.

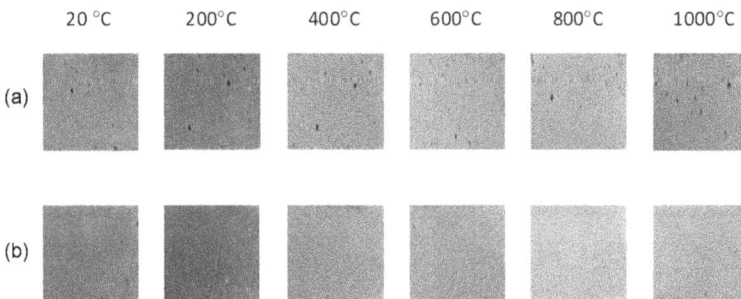

FIGURE 7.1 Geopolymer's surface changes of (a) FA and (b) FA-GGBFS geopolymers upon thermal exposure (Hager et al., 2021).

7.2.2 Mass Loss and Volume Change Analysis

The mass loss values of OPC and the geopolymers upon thermal exposure are listed in Table 7.1. The mass loss trends are likely similar, increasing from 200°C to 800°C and remaining between 800°C and 1,000°C. The increase in mass loss between 200°C and 800°C is related to the evaporation of water during the heating process. In comparison, the mass loss of OPC is higher than that of geopolymer at elevated temperatures, especially at 800°C and 1,000°C. Therefore, geopolymer outperformed OPC at elevated temperatures, which has excellent durability on fire, due to its better structural stability.

The volume changes of the geopolymer and OPC upon thermal exposure are tabulated in Table 7.2. The negative value of volume loss inferred that the geopolymers shrank, while the positive value implied that the geopolymers expanded. The geopolymers experience shrinkage properties at all temperature ranges, except for FA geopolymers, which expand at 1,000°C. Shrinkage is caused by phase conversion within the matrix. In contrast, OPC shrinks little at a temperature lower than 1,000°C but shrinks significantly at a temperature higher than 1,000°C. Thus, the geopolymer with lower shrinkage is preferred over OPC for exposure temperature applications.

7.2.3 Density Analysis

The density loss of different geopolymers and OPC after thermal treatment is displayed in Table 7.3. The density loss trend of the geopolymer and OPC is closely similar when the temperature increases from 200°C to 600°C and is retained up to 1,000°C. The increased density loss is affected by the increased mass loss and volume changes, as aforementioned. It is well known that density is directly proportional

---

TABLE 7.1

Mass Loss of the Geopolymers and OPC

References	Sample	Mass Loss (%)				
		200°C	400°C	600°C	800°C	1,000°C
	OPC	10.10	13.40	17.20	23.70	24.20
	MK geopolymers	13.60	15.00	16.00	16.40	16.50
(Burciaga-Díaz and Escalante-García, 2017)	GGBFS geopolymers	13.20	16.10	17.90	18.60	18.70
	GGBFS-MK geopolymers	10.90	13.10	14.40	15.00	15.10
	FA geopolymers	5.80	8.00	9.80	10.30	10.60
(Hui-Teng et al., 2022)	FA-LFS geopolymers	5.10	8.00	10.80	11.70	11.60

TABLE 7.2

Volume Changes of Geopolymers and OPC

Reference	Sample	Volume Loss (%)					
		200°C	400°C	600°C	800°C	1,000°C	1,200°C
	OPC	–	–	1.30	2.50	6.30	32.50
	MK geopolymers	–	–	12.50	16.30	17.50	30.00
Burciaga-Díaz and Escalante-García (2017)	GGBFS geopolymers	–	–	6.00	7.50	15.00	18.80
	GGBFS-MK geopolymers	–	–	8.00	12.00	34.00	–
	FA geopolymers	2.30	5.30	6.00	7.00	19.50	–
Hui-Teng et al. (2022)	FA-LFS geopolymers	2.30	5.00	5.90	6.60	6.80	–

TABLE 7.3

Density Loss of the Geopolymer and OPC

References	Sample	Density Loss (%)				
		200°C	400°C	600°C	800°C	1,000°C
Boquera et al. (2021)	OPC	7.90	13.20	15.80	15.80	–
Hager et al. (2021)	FA geopolymer	7.20	8.10	8.30	8.10	8.80
	FA-GGBFS geopolymer	6.20	7.40	9.50	8.70	9.40
Hui-Teng et al. (2022)	FA-LFS geopolymer	2.90	3.20	5.20	5.50	5.20
Luo et al. (2022)	FA-GGBFS geopolymer	1.70	1.80	4.80	4.20	–

to mass and inversely proportional to volume. As expected, OPC with higher density loss showed a weaker structure at elevated temperatures when compared to the geopolymer.

7.3 MECHANICAL PROPERTIES

The mechanical evolution includes compressive strength changes and flexural strength changes, which will be discussed in this section. The mechanical evolution of the geopolymer during thermal exposure plays an important role in evaluating the performance of the geopolymer at elevated temperatures.

7.3.1 COMPRESSIVE STRENGTH ANALYSIS

Payakaniti et al. (2020) prepared a high-calcium FA geopolymer in order to evaluate the effect of elevated temperatures on its compressive strength. The geopolymer's compressive strength increased from 57 to 71 MPa at 200°C as a result of the further geopolymerization. As the temperature rose, the compressive strength decreased. At

400°C, 600°C, 800°C, 1,000°C, and 1,200°C, the compressive strengths were 65.00, 40.00, 30.00, 31.50, and 7.00 MPa, respectively. On the other hand, MK geopolymer prepared by Albidah et al. (2021) was exposed to 200°C, 400°C, and 600°C. When the temperature rose from ambient temperature to 200°C, a considerable amount of compressive strength was degraded. After exposure to 200°C, 400°C, and 600°C, the compressive strength varied by roughly 7.0%, 13.0%, and 6%, respectively. The different types of aluminosilicate precursors contribute to the different strength development of the geopolymer during the elevated temperature exposure due to the contents of Al and Si.

Hassan et al. (2020) reported that the compressive strength of FA geopolymer increased by about 33% and 18% at 200°C and 400°C, respectively, compared to unheated FA geopolymer. However, the strength decreased by roughly 18%, 27%, and 63% with a subsequent temperature of 600°C, 800°C, and 1,000°C, respectively. When compared to OPC, the compressive strength was preserved up to 400°C and began to decline at 800°C. Beyond 400°C, the OPC cracked readily, which may be caused by the moisture evaporating quickly from the matrix.

Additionally, Rao and Kumar (2020) investigated the response of FA-GGBFS geopolymer at temperatures of 200°C, 400°C, 600°C, and 800°C. The achieved compressive strength at 200°C was around 40% higher than that of unheated geopolymers. Even if the pattern of decline continued up to 800°C, the variations between 600°C and 800°C were more noticeable. Even though the FA-GGBFS geopolymer's degradation began at 600°C, it still had a higher strength rate of 80% compared to unheated geopolymer. Compressive strength would increase or decrease depending on the heat exposure time. According to Hafez et al. (2019), the FA-GGBFS geopolymer declined in compressive strength by around 11% at 300°C as the exposure time increased from 1 to 2 hours. Similar to OPC, it displayed a 25%–27% reduction in compressive strength from its initial values.

7.3.2 FLEXURAL STRENGTH ANALYSIS

FA-LFS geopolymer, which was subjected to 300°C, 600°C, 900°C, and 1,100°C, was tested by Yong-Sing et al. (2022). The FA-LFS geopolymer's maximum limit of temperature was chosen at 1,100°C, as the FA-LFS geopolymer would melt beyond the temperature. The flexural strength of FA-LFS geopolymer (7.8 MPa) increased as the exposure temperature increased, which was attributed to an exothermic reaction due to the formation of more reaction products during the geopolymerization. The release of moisture within the geopolymer caused a modest drop in flexural strength to 6.90 MPa at 300°C. The escaping water molecule increased porosity, weakening the structure and lowering its strength. A significant rise in flexural strength (24.10 MPa) was shown when the temperature increased to 1,100°C due to the densification effect of the geopolymer. The formation of refractory phases and surface solidification filled the pores and cracks, resulting in greater flexural strength.

After being exposed to 100°C, 300°C, 500°C, and 700°C, Zhang et al. (2014) analyzed the flexural strength of MK-FA geopolymer. The geopolymer's flexural strength showed a modest rise at 100°C, then a decline in strength from 300°C to 700°C. At 100°C, OPC showed an increase similar to that of the geopolymer. However, at room temperature, the strength of the geopolymer was greater than that of OPC, but it decreased when exposed to higher temperatures. Due to the formation of internal microstructure cracks, such as the propagation of cracks and the generation of pores at elevated temperatures, the strength degraded as a result of the temperatures.

7.4   MATERIAL CHARACTERIZATION

The concern about the effect of elevated temperature exposure on the geopolymer, especially according to microstructure and phase changes, is explained. Considering that these methods can effectively enlighten the intervening geopolymer under elevated temperatures due to sintering and partial melting.

7.4.1   MICROSTRUCTURAL ANALYSIS

The microstructure changes of FA and FA-LFS geopolymers before and after heat treatment were performed by Hui-Teng et al. (2022), as shown in Figure 7.2. Increasing temperatures loosen the geopolymer with increased pores and cracks up to 600°C, thereby weakening the structure. A similar condition was observed in FA geopolymers up to 1,000°C, resulting in poor mechanical strength. The FA-LFS geopolymer had a dense microstructure between 800°C and 1,000°C owing to the occurrence of the densification effect. In comparison, the participation of LFS improved the microstructure compactness of the FA geopolymer, causing the FA-LFS geopolymer to achieve excellent mechanical strength.

Geopolymers based on red mud were tested by Lemougna et al. (2017) at 60°C, 300°C, and 800°C. They found that the microstructure connectivity of the matrix increased with increasing temperatures up to 800°C, which is attributed to the sintering effect, and resulted in improved mechanical strength. Conversely, Burciaga-Díaz and Escalante-García (2017) revealed that the microstructure of MK, GGBFS, and MK-GGBFS geopolymers, as well as OPC, became looser as the temperature increased from 20°C to 600°C, 1,000°C, and 1,200°C. The poorer microstructure was caused by the occurrence of a thermal mismatch between the new phase and the matrix itself, as well as the increased pores and cracks, thereby decreasing the mechanical strength at elevated temperatures.

7.4.2   PHASE CHANGE ANALYSIS

The phase changes of varying geopolymers and OPC upon heat treatment are done by Burciaga-Díaz and Escalante-García (2017), as demonstrated in Figure 7.3. At room temperature, the amorphous nature (GGBFS geopolymer: 25–35° 2θ; MK geopolymer and MK-GGBFS geopolymer: 15–35° 2θ) was present in the geopolymer but absent in OPC. The increase in temperature reduced and converted the amorphous hump of GGBFS (Figure 7.3a) and MK-GGBFS geopolymers (Figure 7.3c) into crystalline phases, while the hump was retained in MK geopolymer (Figure 7.3b) at elevated temperatures. However, in general, the crystalline phase of the geopolymers, as well as OPC, changed, increased, reduced, or decomposed as the

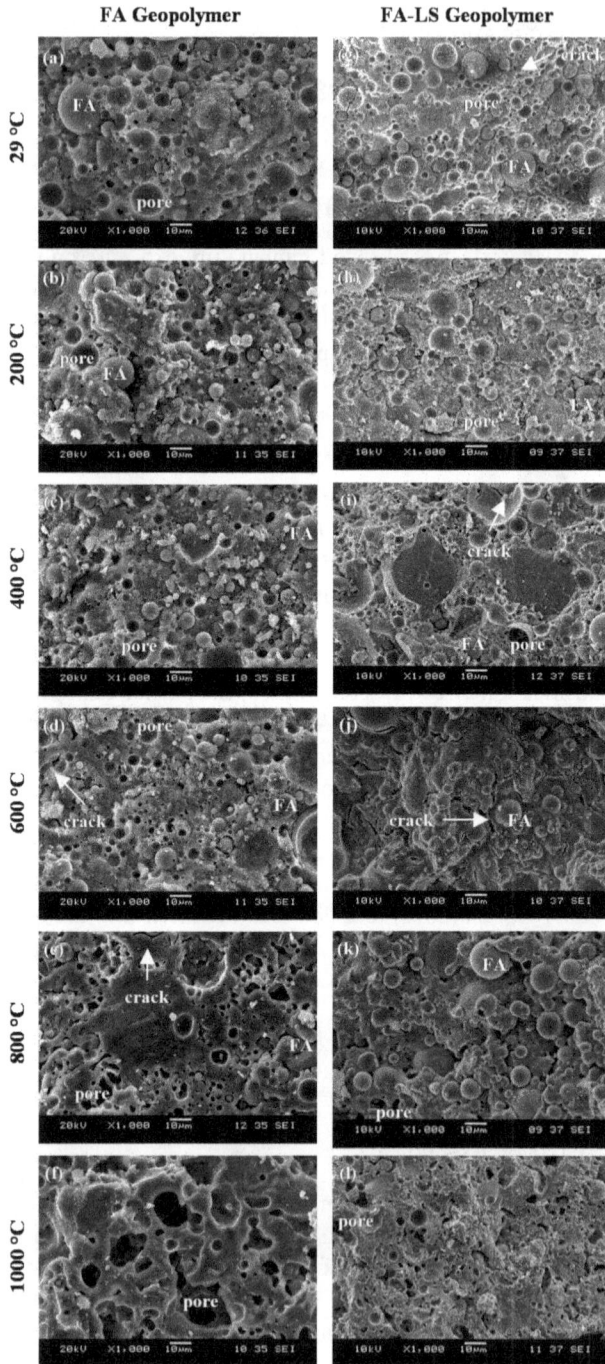

FIGURE 7.2 Microstructure of FA and FA-LFS geopolymers upon thermal exposure (Hui-Teng et al., 2022).

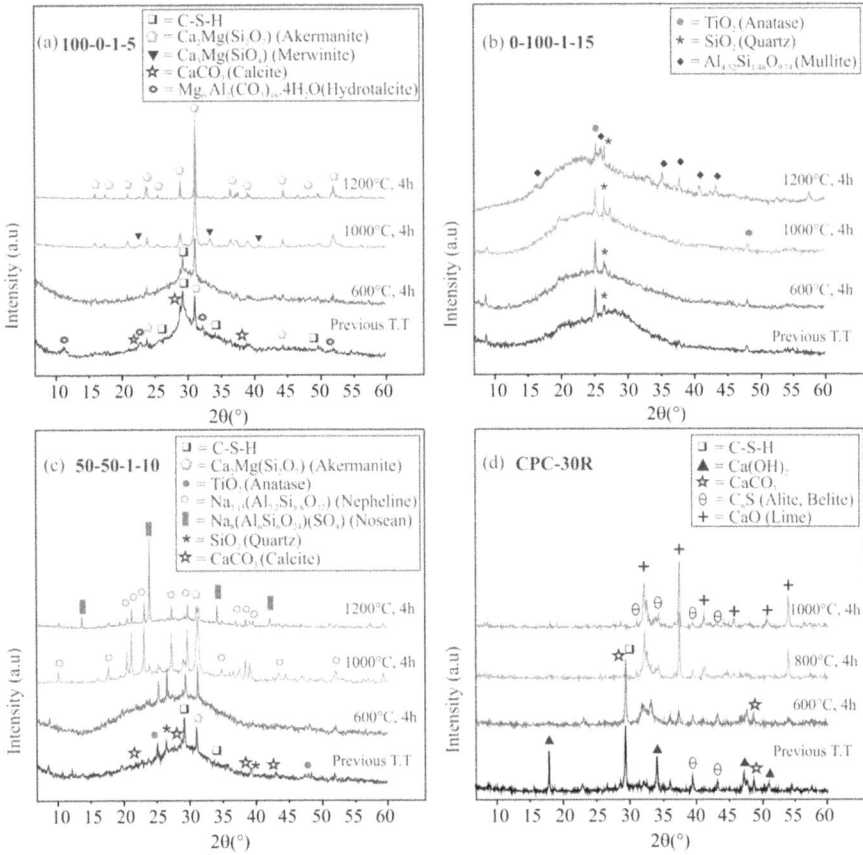

FIGURE 7.3 Phase changes of (a) GGBFS geopolymer, (b) MK geopolymer, (c) MK-GGBFS geopolymer, and (d) OPC upon thermal exposure (Burciaga-Díaz and Escalante-García, 2017).

temperature increased. These phase transformations are the main factors that damaged the internal structure of the binder, which should crack, pores, or even spall, thereby resulting in severe mechanical strength at elevated temperatures. A similar statement was made by Hui-Teng et al. (2022) and Niklioć et al. (2016), who produced FA and FA-EAFS geopolymers, respectively.

7.5 CONCLUSIONS

The demand for sustainable and eco-friendly construction materials has increased significantly in recent years due to growing concern over the impact of construction activities on the environment. Geopolymers, which are synthesized by the geopolymerization process, have emerged as a promising alternative to traditional building materials such as OPC. One of the major advantages of geopolymers is their excellent fire resistance, which makes them ideal for use in high-risk environments where fire safety is a priority.

The excellent fire-resistant properties of geopolymers are attributed to their unique molecular structure, which is formed during the geopolymerization process. Geopolymers have a highly cross-linked structure, which gives them high strength and durability and makes them highly resistant to fire. The global scarcity of materials with improved thermophysical properties has also heightened the need for sustainable construction materials, making geopolymers a viable solution for the industry.

In conclusion, the fire-resistant properties of geopolymers make them a viable and practical solution for the construction industry. The use of geopolymers can enhance the safety of construction projects and significantly reduce the environmental impact of the industry. As a result, it is recommended that geopolymerization methods be adopted more widely in the construction industry, especially in areas where fire safety is a priority.

REFERENCES

Abd El Hafez, R. D., Ahmad, A. R. M., Khafaga, M. A., & Refaie, F. A. Z. (2019). Effect of exposure to elevated temperatures on geopolymer concrete properties. *International Journal of Civil Engineering and Technology*, *10*(10), 448–461.

Albidah, A., Alghannam, M., Abbas, H., Almusallam, T., & Al-Salloum, Y. (2021). Characteristics of metakaolin-based geopolymer concrete for different mix design parameters. *Journal of Materials Research and Technology*, *10*, 84–98. doi:10.1016/j.jmrt.2020.11.104

Boquera, L., Castro, J. R., Pisello, A. L., Fabiani, C., D'Alessandro, A., Ubertini, F., & Cabeza, L. F. (2021). Thermal and mechanical performance of cement paste under high temperature thermal cycles. *Solar Energy Materials and Solar Cells*, *231*, 111333. doi:10.1016/j.solmat.2021.111333

Burciaga-Díaz, O., & Escalante-García, J. I. (2017). Comparative performance of alkali activated slag/metakaolin cement pastes exposed to high temperatures. *Cement and Concrete Composites*, *84*, 157–166. doi:10.1016/j.cemconcomp.2017.09.007

Davidovits, J. (2020). *Geopolymer Chemistry and Applications* (5th edn.). Saint-Quentin, France.

Hager, I., Sitarz, M., & Mróz, K. (2021). Fly-ash based geopolymer mortar for high-temperature application - Effect of slag addition. *Journal of Cleaner Production*, *316*. doi:10.1016/j.jclepro.2021.128168

Hassan, A., Arif, M., & Shariq, M. (2020). Mechanical behaviour and microstructural investigation of geopolymer concrete after exposure to elevated temperatures. *Arabian Journal for Science and Engineering*, *45*(5), 3843–3861. doi:10.1007/s13369-019-04269-9

Hui-Teng, N., Cheng-Yong, H., Yun-Ming, L., Abdullah, M. M. A. B., Pakawanit, P., Bayuaji, R., Yong-Sing, N., Zulkifly, K. B., Wan-En, O., Yong-Jie, H., & Shee-Ween, O. (2022). Comparison of thermal performance between fly ash geopolymer and fly ash-ladle furnace slag geopolymer. *Journal of Non-Crystalline Solids*, *585*, 121527. doi:10.1016/j.jnoncrysol.2022.121527

Krishna Rao, A., & Rupesh Kumar, D. (2020). Comparative study on the behaviour of GPC using silica fume and fly ash with GGBS exposed to elevated temperature and ambient curing conditions. *Materials Today: Proceedings*, *27*, 1833–1837. doi:10.1016/j.matpr.2020.03.789

Lemougna, P. N., Wang, K. T., Tang, Q., & Cui, X. M. (2017). Synthesis and characterization of low temperature (<800°C) ceramics from red mud geopolymer precursor. *Construction and Building Materials*, *131*, 564–573. doi:10.1016/j.conbuildmat.2016.11.108

Luo, Y., Li, S. H., Klima, K. M., Brouwers, H. J. H., & Yu, Q. (2022). Degradation mechanism of hybrid fly ash/slag based geopolymers exposed to elevated temperatures. *Cement and Concrete Research*, *151*. doi:10.1016/j.cemconres.2021.106649

Niklioć, I., Marković, S., Janković - Častvan, I., Radmilović, V. V., Karanović, L., Babić, B., & Radmilović, V. R. (2016). Modification of mechanical and thermal properties of fly ash-based geopolymer by the incorporation of steel slag. *Materials Letters*, *176*, 301–305. doi:10.1016/j.matlet.2016.04.121

Payakaniti, P., Chuewangkam, N., Yensano, R., Pinitsoontorn, S., & Chindaprasirt, P. (2020). Changes in compressive strength, microstructure and magnetic properties of a high-calcium fly ash geopolymer subjected to high temperatures. *Construction and Building Materials*, *265*, 120650. doi:10.1016/j.conbuildmat.2020.120650

Yong-Sing, N., Yun-Ming, L., Cheng-Yong, H., Abdullah, M. M. A. B., Pakawanit, P., Chan, L. W. L., Hui-Teng, N., Shee-Ween, O., Wan-En, O. & Yong-Jie, H. (2022). Thin fly ash/ladle furnace slag geopolymer: Effect of elevated temperature exposure on flexural properties and morphological characteristics. *Ceramics International*, *48*(12), 16562–16575. doi:10.1016/j.ceramint.2022.02.201

Zhang, H. Y., Kodur, V., Qi, S. L., Cao, L., & Wu, B. (2014). Development of metakaolin-fly ash based geopolymers for fire resistance applications. *Construction and Building Materials*, *55*, 38–45. doi:10.1016/j.conbuildmat.2014.01.040

8 Geopolymers as Coating Material

Liyana Jamaludin, Mohd Mustafa Al Bakri Abdullah, and Shamala A. P. Ramasamy
Universiti Malaysia Perlis

Petrica Vizureanu
Gheorghe Asachi Technical University

8.1 INTRODUCTION

Existing cementitious coating or repair material is mainly ordinary Portland cement (OPC)-based for almost all available cementitious coating. Apart from contributing to CO_2 emissions, the harmful and hazardous materials used in the pre- and post-production of this inorganic coating might also volatilize into the atmosphere. Final applications of cementitious coatings require dependable adhesive strength and outstanding mechanical qualities that are unaffected by sample age. Materials called geopolymers are created when an inorganic substance like kaolin, fly ash, powdered granulated blast slag, or rice husk reacts with an alkaline activator (Ikmal et al., 2020). Geopolymers have impressive fire resistance, the capacity to encapsulate hazardous waste, and are inexpensive, as reported by Davidovits (1994). Great research is needed to fully utilize their usage as a coating material. Geopolymer coating provides additional surface protection and aids in maintaining it in like-new condition with relatively little upkeep and an easily cleanable surface. Geopolymer has been a popular alternative for the last three decades to work on replacing OPC-based applications. Geopolymer coating as a coating material would offer corrosion resistance, protect structural integrity, and prevent wear to its substrates (Ana María et al., 2018). Applying coatings has been proven efficient in inhibiting the entrance of harmful ions, thus reducing the resultant deterioration (Wang et al., 2022). Additionally, geopolymer coatings have excellent potential for surface protection, enhanced structural durability, and suitability for application in high-temperature exposure situations. One possible alternative to improving the strength of ceramic is to improve its coating properties. The most efficient way to reduce spalling and cracking of surfaces when exposed to high temperatures is to prevent liquid ingress into ceramic, thus preventing the ingress of chemicals and crack deformation (Safiuddin et al., 2017).

Protective coating is being used as an alternative to seal and protect the concrete surface, but the most common coatings are based on organic matrices and epoxy

DOI: 10.1201/9781003390190-8

(Papakonstantinou and Balaguru, 2017). Geopolymer coatings have shown substantial attention as an alternative material for protective coatings on concrete surfaces (Zhu, 2020). A novel kaolin and fly ash-based geopolymer coating is a significant invention for our planet when taking into account the commercial viability and long-term consumption of non-metallic cementitious coatings and repair materials. The properties and durability of geopolymeric coatings were measured based on the mechanical properties and morphology analysis present in the coating, which are adhesion strength, water absorption, and phase analysis.

8.2 MIX DESIGN PARAMETERS OF GEOPOLYMER COATING PASTE

8.2.1 INFLUENCE OF SOLID/LIQUID (S/L) OF FLY ASH GEOPOLYMER PASTE

The fly ash geopolymer coating paste was subjected to a solid/liquid (S/L) ratio range of 1.0–2.5 for a 24h curing period. With an increase in fly ash content, the compressive strength of fly ash-based geopolymer paste increased simultaneously. The S/L ratio has an impact on the mechanical strength development of geopolymer paste. From the data obtained, the S/L ratio was 1.0, which contributed to the low strength of the geopolymer sample due to the high content of alkaline activator and the increasing production of excessive OH- left in the geopolymer system. The concentration and composition of alkaline activators play an important role in developing the mechanical strength of geopolymer paste (Adrian et al., 2019). Figure 8.1 shows the compressive strength of geopolymer paste with S/L ratios of 1.0, 1.5, 2.0, and 2.5. The maximum strength for an S/L ratio of 2.0 is 52 MPa. As the S/L ratio rises to 2.0, compressive strength tends to increase until it drops to 5 MPa at S/L ratio 2.5. The development of mechanical strength in geopolymers is affected by the ratio S/L. With the exception of the ratio at 2.5, which substantially dropped at

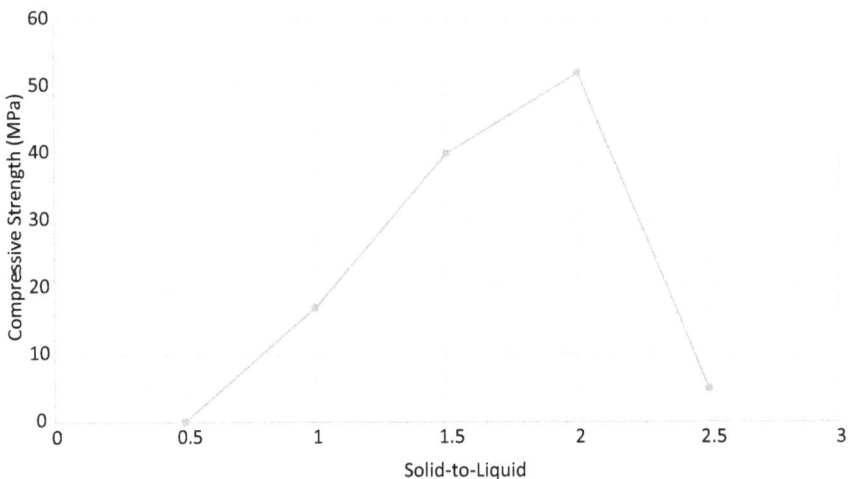

FIGURE 8.1 Compressive strength of fly ash geopolymer paste with a solid/liquid ratio varying from 1.0, 1.5, 2.0, and 2.5.

5 MPa due to the limited workability of the geopolymer paste, the strength findings rise when the S/L ratio is increased. The optimum S/L ratio was obtained at 2.0. Thus, this ratio was used to apply coating material to the specimen. The rate of geopolymerization is affected by the initial solid content. Increasing the S/L ratio leads to a larger formation of alkaline aluminosilicate as the reactant species are dissolved more (Yong Sing et al., 2020).

8.2.2   INFLUENCE OF S/L OF KAOLIN GEOPOLYMER PASTE

Figure 8.2 shows the compression strength of kaolin geopolymers with varying S/L ratios cured at 70°C for 24 hours. The data from the figure show that the compressive strength of kaolin geopolymers increased linearly with increasing S/L ratios for samples. The early strength increased with increasing S/L ratios, from 0.7 to 1.1. The highest compressive strength of 2.45 MPa achieved after 7 days was for a sample with a kaolin-to-alkaline activator ratio of 1.1, while a sample with an S/L ratio of 0.7 showed the lowest compressive strength of 0.45 MPa. It also showed that, over time, the compressive strength of kaolin geopolymers increased. Samples after 28 days and 90 days also showed the highest compressive strengths of 2.65 and 4.44 MPa for an S/L ratio of 1.1. Regardless of the strength gain of the sample with an S/L ratio of 1.1, the poor workability and difficulty in compacting made it not possible to be taken as the optimum formulation. With good workability and a nearly identical strength improvement of 4.15 MPa after 90 days, the sample with an S/L ratio of 0.9 was the next most promising formulation. A highly workable sample with an S/L ratio of 0.7, however, exhibits a low increase in strength even after 90 days (0.78 MPa).

The activating solution and the reacting materials made better contact when the S/L ratio was raised, which led to an increase in the observed compressive strength.

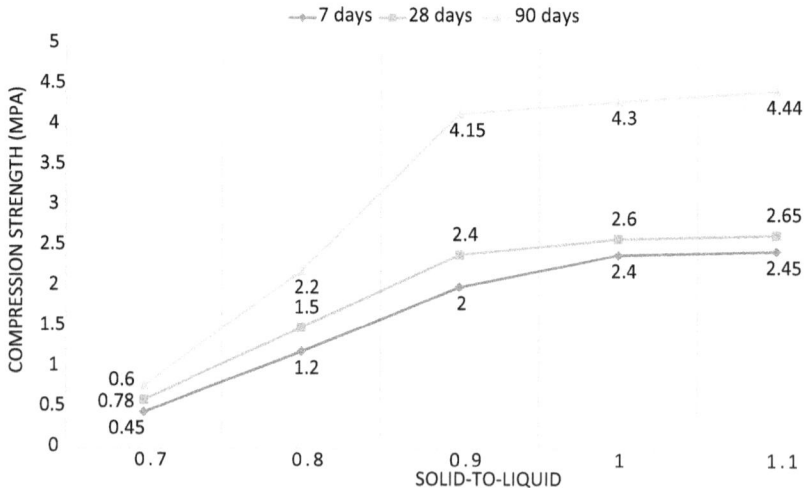

FIGURE 8.2   Compression strength of kaolin geopolymer paste with a solid/liquid ratio varying from 0.7, 0.8, 0.9, 1.0, and 1.1.

Strength growth for samples with lower S/L ratios of 0.7 and 0.8 was weak and poor. Despite having excellent workability, samples with a low kaolin-to-alkaline activator ratio showed almost no increment due to the high amount of activating solution that hindered geopolymerization process. An excess amount of alkaline solution is available, but a lack of kaolin makes the contact between activation solutions and reacting raw materials limited. The overall compressive strength of kaolin geopolymers showed the highest strength for S/L of 1.1, but its low workability makes it not suitable as an optimum S/L ratio. Considering the preparation period of kaolin geopolymers, an S/L ratio of 0.9 is fixed as an optimum as it provides the highest strength with good workability as well. Therefore, the optimum S/L ratio of 0.9 for kaolin geopolymer paste synthesis was carried forward as coating materials onto substrates.

8.3 PREPARATION OF FLY ASH-BASED GEOPOLYMER AT CERAMIC INDUSTRY SUBSTRATE AND KAOLIN-BASED GEOPOLYMER COAT FOR LUMBER-GRADE WOOD

Fly ash that has been sieved, an alkaline activator solution with an S/L ratio of 2.0, and a sodium hydroxide to sodium silicate ratio of 2.5 at a natrium hydroxide molarity of 10 M were used. The alkaline solution was added to the powdered fly ash and stirred for 5 minutes to generate a homogeneous slurry. Fly ash geopolymer slurry paste is a combination used for coating applications. Industrial ceramic substrate samples of 100 mm × 50 mm were coated with geopolymer ceramic coating materials using a brushing technique. The coating's thickness ranged from 0.3 to 0.5 mm. Ceramic samples were brush-coated and then allowed to cure for 24 hours at room temperature. Thermally treated substrates must undergo testing after curing.

For kaolin geopolymer coating, the kaolin was mixed with alkaline activator solution at the desired ratio, ranging from 0.7 to 1.1. The mixture was then stirred via a mechanical stirrer at a speed of 20 rpm for 5 minutes until a homogenous slurry was obtained. Upon mixing, kaolin-based geopolymer paste is obtained, and to achieve an evenly coated layer of kaolin geopolymer coating, a customized, self-invented sliding panel was used for the coating process. Lumber-grade wood substrates were placed at the base of the panel. After which, geopolymer paste was slid evenly onto LG wood, as demonstrated in Figure 8.3. Once it is coated evenly, excess paste from the edge is wiped off, and the sample is allowed to dry for about 1 hour on the panel itself. This is to ensure that the coating layer is not disturbed in its gel phase. After the first hour, using fine-tip forceps, the coated LG wood substrates were carefully removed from the sliding panel and placed on a tray. Samples are ready for the next stage which is the curing process.

A curing temperature of 70°C for 24 hours is used. After curing in oven, samples were removed, allowed to cool to room temperature, and sealed in thin plastic bags to prevent dust and impurities from entering the environment until the day of testing.

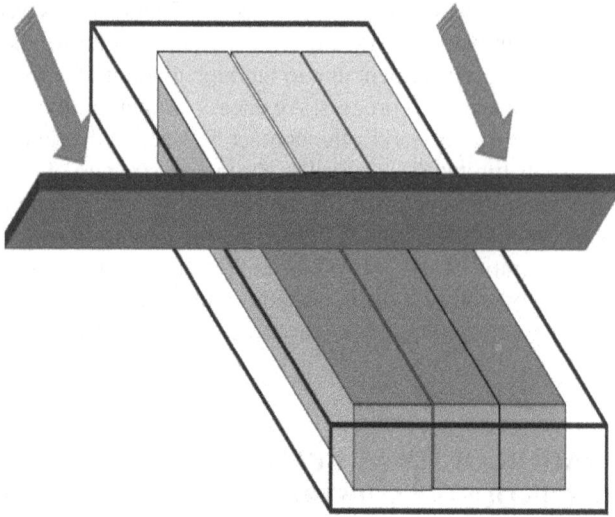

FIGURE 8.3 Demonstration of coating process of kaolin geopolymer on LG wood substrate.

8.4 PHYSICAL AND MECHANICAL PROPERTIES OF GEOPOLYMER COATING

8.4.1 ADHESION STRENGTH ANALYSIS

Figure 8.4 shows the condition of the geopolymer ceramic coating with different thermal conditions applied. The adhesion strength was evaluated after 3 days. The results reveal that the unsintered geopolymer coating's adhesive strength remains constant at 1.2 MPa. Up to 1,500°C of sintering temperature, the adhesion strength grew stronger. The greatest strength was 4.8 MPa for 3 days at 1,500°C. At 1,600°C, the strength dropped dramatically to 0.3 MPa.

According to the results, the link between matrix and geopolymer and the temperature have an impact on the adhesion strength of coating surface substrates. This problem happens as a result of the coating material's poor ability to disperse over the excessively high sintered substrate surface. The adhesion strength results correlate with previous research from Klima et al. (2022), who reported that geopolymers were stable up to 1,300°C, making them perfect candidates to be used for thermal protection of concrete structures. Such applications may include, but are not limited to, fire-prone construction buildings, nuclear plants, and other structures subjected to high-temperature environments. High fire resistance of geopolymer ceramic enhanced the heat resistance of performance structure.

Figure 8.5 represents the adhesion strength of kaolin geopolymer-coated lumber wood with various S/L ratios cured at 70°C for 24 hours. These samples were analyzed and tested after 7, 28, and 90 days, respectively. Kaolin geopolymer-coated lumber wood with an S/L ratio of 1.1 showed the highest bonding strength of 4.5 MPa for samples tested after 7 days to study early bonding strength. An almost

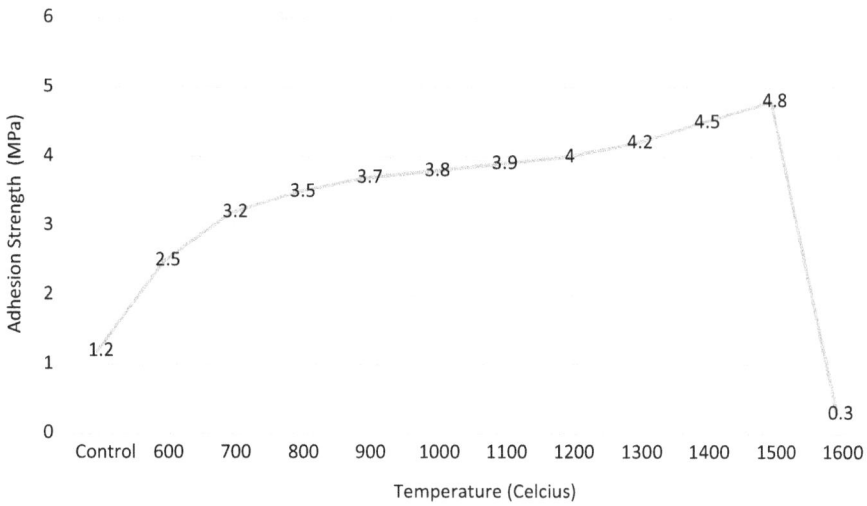

FIGURE 8.4 Adhesion strength of geopolymer ceramic coating with different sintering temperatures.

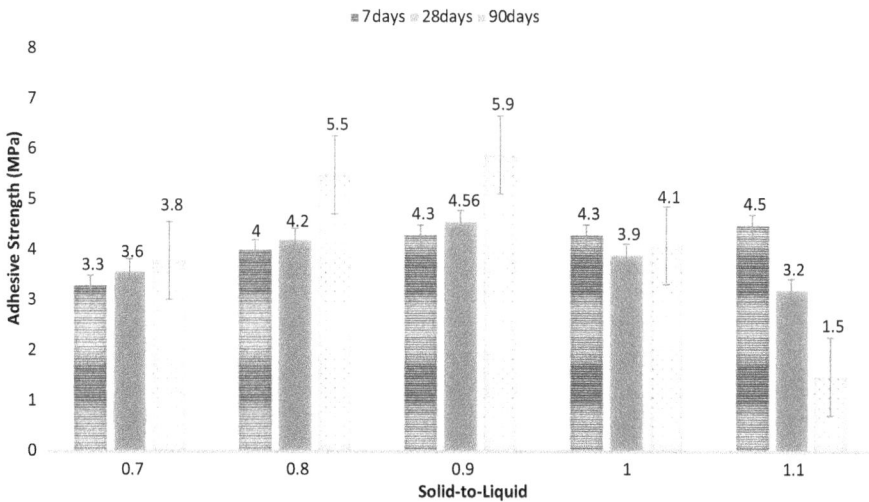

FIGURE 8.5 Adhesive strength of kaolin geopolymers coated lumber wood with varying solids-to-liquid (S/L) ratios tested after 7, 28, and 90 days.

linear trend is observed (Figure 8.5) in how the S/L ratio influences the adhesive strength of kaolin geopolymer-coated lumber woods tested after 7 days. However, as samples were tested over time, kaolin geopolymer-coated lumber wood with an S/L ratio of 0.9 showed a progressively increasing and ambitious adhesive strength of 4.56 MPa (28 days) and 5.90 MPa (90 days). In contrast to the bonding strength given upon 7 days of testing, kaolin geopolymers coated lumber wood with an S/L ratio of 1.1 tested after 90 days provided comparatively the lowest adhesive strength of 1.50 MPa. It is also observed that this sample readily peels upon minor loading on contact.

A sufficient quantity of water content in the geopolymer coating results in better binder formation and superior workability, which boost adhesive strength in samples with an S/L ratio of 0.9. Though sufficient water is required to obtain a crack-free coating layer, excessive residual water existing in kaolin-based geopolymers coated lumber wood (S/L) ratios of 0.7–0.8 can weaken its properties.

Kaolin geopolymer-coated lumber wood samples with an S/L ratio of 1.1 showed the worst progressive degradation of adhesive strength over sample age. Bonding properties of samples drastically dropped due to the excessive amount of unreacted kaolin particles that disrupted the geopolymer structure over time. Development of crystalline phase (zeolites) of geopolymer also caused the sample to become more brittle and porous, which led to extremely poor bonding properties. These findings are in agreement with previous researcher Nikolov et al. (2017). Increasing the S/L ratio above 0.9 induces more micro-cracks on the coating surface, thus leading to crack penetration over time that weakens the adhesive strength. This is because a high S/L ratio causes a high risk of segregation and porous surfaces as well (Haleh et al., 2020).

Based on the S/L ratio investigated in our work, it can be asserted that raising the S/L ratio improves the early development of adhesion strength in geopolymer-coated wood based on kaolin. The finest bonding capabilities, however, are promised by kaolin geopolymer-coated timber wood with an S/L ratio of 0.9 when workability, processing step, and long-term adhesive strength are taken into consideration.

8.4.2 WATER ABSORPTION ANALYSIS

The water absorption test was conducted to investigate the water-resistance properties of geopolymer ceramic coatings. The testing was conducted after 3 days of coating, and the results are shown in Figure 8.6. The percentage of water absorption of 8% was recorded for the unsintered geopolymer coating. The water absorption test conducted for 24 hours shows a decreasing pattern with the increasing sintering temperature. The results of water absorption at 1,500°C show the lowest percentage of 2%. When the sintering temperature reached 1,600°C, the water absorption started to increase up by to 10%.

The surface condition of coating layer influences the water absorption in geopolymer coatings. Overall, when temperature increased up to 1,500°C, the water absorption reduced as a result of the sintering process, which minimized inter-particle voids. Lower water absorption also showed greater resistance to concentrations of water. The increasing water absorption at 1,600°C was attributed to the heterogeneous melting of the coating surface, which influenced the crack formation and high absorption of water in the substrate.

Figure 8.7 displays the water absorption percentage (WA%) of kaolin geopolymers with varying S/L ratios from 0.7 to 1.1. The water absorption test was conducted to investigate the water-resistance properties of kaolin geopolymer paste. The immersion

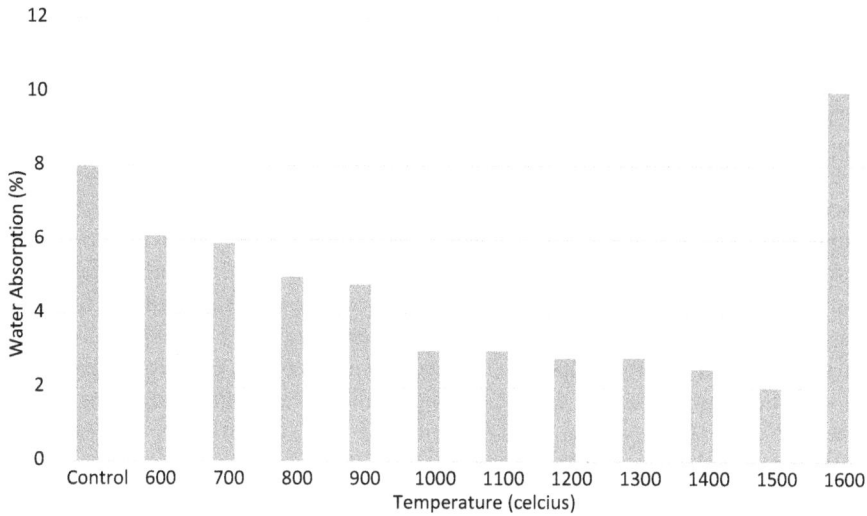

FIGURE 8.6 Percentage water absorption of geopolymer ceramic coatings for different sintering temperatures.

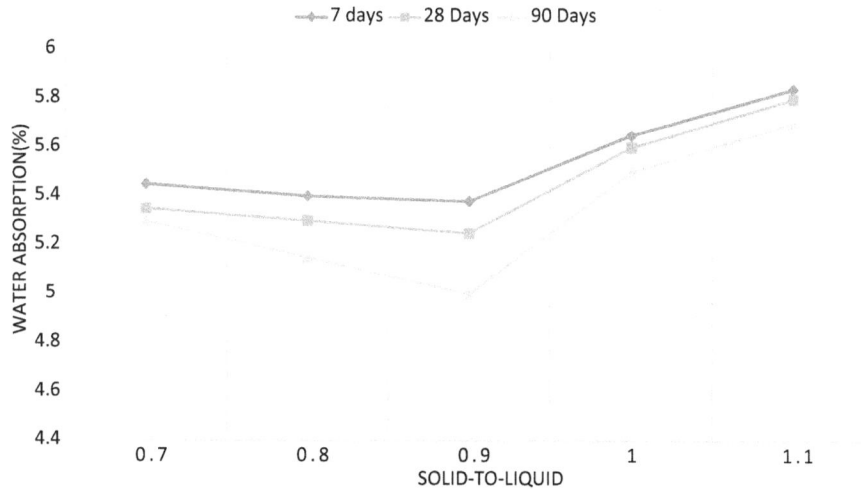

FIGURE 8.7 Water absorption of kaolin geopolymer paste with varying solids-to-liquid (S/L) ratios from 0.7 to 1.1.

test was conducted for 24 hours on samples at 7, 28, and 90 days of age. The water-resistance test conducted also showed impressive early results for all samples. The WA% for all samples was within the range of 5.00%–6.00%. The lowest WA% of kaolin geopolymers is 4.98 given by sample with an S/L ratio of 0.9 tested after 90 days. On the other hand, the highest WA% value of 5.84 was observed for samples with an S/L ratio of 1.1 tested after 7 days. Figure 8.7 clearly indicates that the WA% of kaolin geopolymer with an S/L ratio of 0.9 drops drastically after 90 days as compared to other kaolin geopolymers with similar S/L ratios. Kaolin geopolymer with an S/L ratio of 0.9 consistently exhibited the lowest WA% for all sample ages. When the S/L ratio of kaolin geopolymer was further increased, WA% also increased regardless of sample age. When Figure 8.7 is analyzed in regard to sample age, all kaolin geopolymers showed decreasing WA% with increasing sample age.

Lower WA% of kaolin geopolymers with an S/L ratio of 0.9 shows higher resistance to water penetration, which also ensures lesser environmental damages. This also proves that kaolin geopolymers with an S/L ratio of 0.9 achieved the most complete geopolymerization process and led to a more stable geopolymer structure as compared to other kaolin geopolymer formulations. WA% decreased linearly with sample age, supporting the theory that geopolymers have better properties over time. This showed that kaolin geopolymer formed a more compact structure that minimized the existing pores and established higher densification that disregards further penetration of water. A similar trend of decreasing WA% over the geopolymer testing period was found by researcher Castillo et al. (2021). Overall, the water absorption in each case was not significantly higher, and all samples gave results below 6% water absorption, which is rather impeccable. This is because, as per the ASTM C90 standard specifications, kaolin achieved WA% values.

8.5 MICROSTRUCTURE PROPERTIES OF GEOPOLYMER COATING

8.5.1 Scanning Electron Microscope Analysis

Microstructure of unsintered is shown in Figure 8.8, and the microstructure of sintering at 1,500°C as the optimum temperature of geopolymer ceramic coating is displayed in Figure 8.9. The spherical form of the fly ash still existed in the microstructure of the unsintered geopolymer ceramic covering, and the surface area

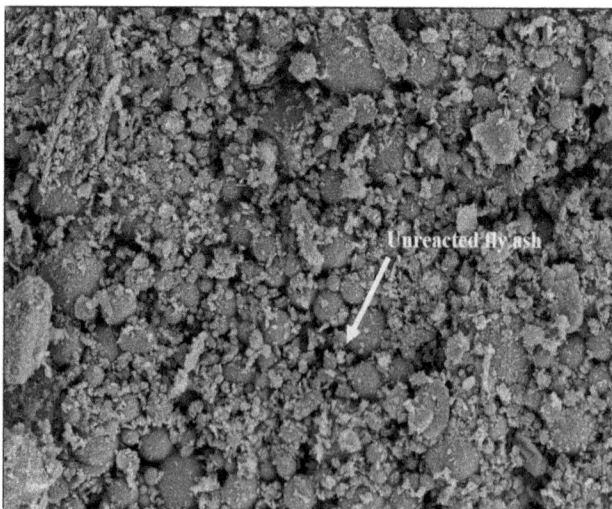

FIGURE 8.8   Microstructure of unsintered geopolymer ceramic coating.

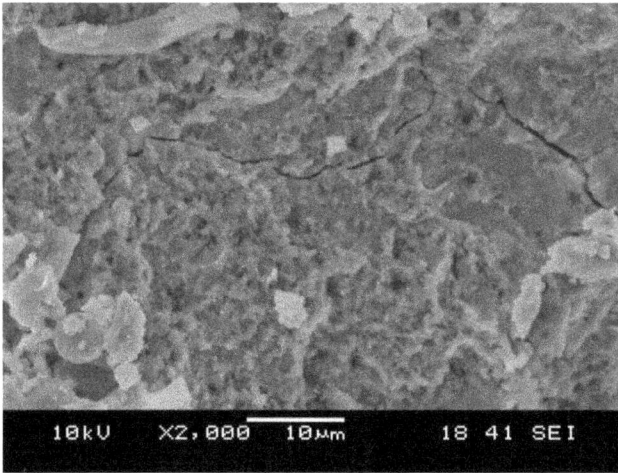

FIGURE 8.9 Microstructure of sintered geopolymer ceramic coating.

displayed unreacted fly ash. For the sintered geopolymer ceramic coating at 1,500°C, the coating surface becomes homogeneous with more intervening materials compared to the unsintered coating. The images mostly contained large amounts of reacting raw materials in the system. The existence of micro-cracks was clearly seen along the interface between the geopolymer coating layer.

The spherical shape of fly ash observed in the unsintered geopolymer ceramic coating was due to unreacted particles between fly ash and alkaline activator. The microstructure of geopolymer ceramic coating clearly shows a significant change in structure with the sintering temperature, which simultaneously gives a higher value of strength. The images of geopolymer ceramic coating obviously showed the fly ash has been stimulated by the alkaline activator solution with alters at the edges of the irregular plate-like particles (Fifinatasha et al., 2016). The form of geopolymer matrix as a compacted structure proved that the geopolymerization reaction occurred at almost all parts of the fly ash particle.

Figure 8.10 displays the microimages of kaolin geopolymer-coated lumber wood with an S/L ratio of (a) 0.7, (b) 0.8, (c) 0.9, (d) 1.0, and (e) 1.1 tested after 90 days. Progressive changes that took place over the sample's age were observed through these microimages.

Interfacial layer with the lowest S/L ratio, after 90 days, gave a coating layer that was rather thin (Figure 8.10a). This is because of the rapid loss of moisture from the excessive, unreacted alkaline activator solution that was merely existing in the geopolymer matrix, thus proving that the S/L ratio is a crucial parameter to produce geopolymer coating with uniform coating thickness and layer. Evaporation of moisture from the top coating layer also developed surface cracks over time. Crack propagation toward inner layers over time will deteriorate the bonding properties of kaolin geopolymer-coated lumber woods. A sample with an S/L ratio of 0.8 showed a more stable interfacial layer image as compared to a sample with an S/L ratio of 0.7 (Figure 8.10b). Open pores that lead to micro-crack propagation were also observed

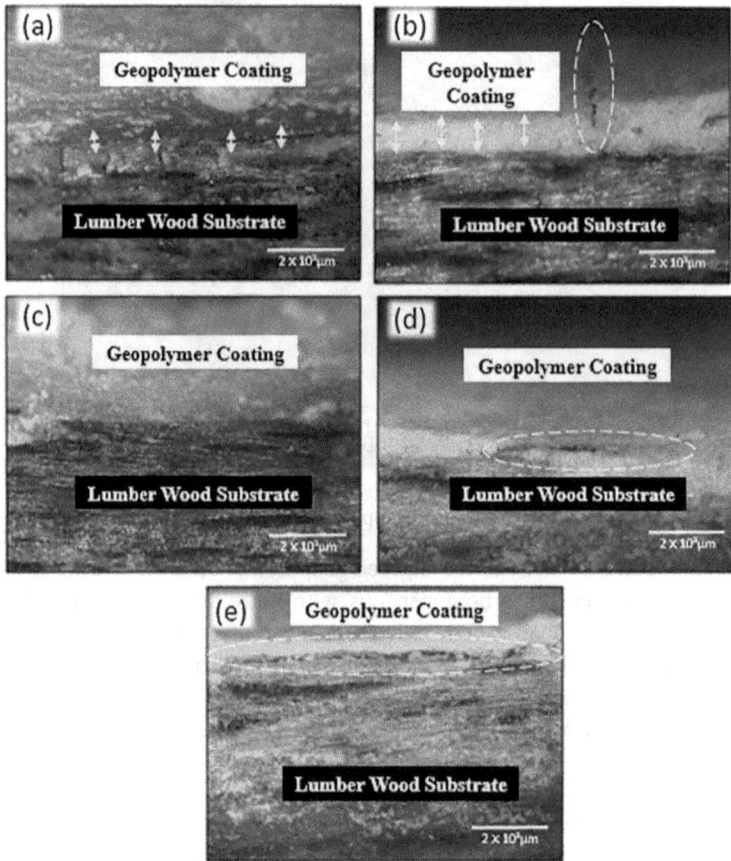

FIGURE 8.10 Micro images of kaolin geopolymer-coated lumber wood with a solids-to-liquid (S/L) ratio of (**a**) 0.7, (**b**) 0.8, (**c**) 0.9, (**d**) 1.0, and (**e**) 1.1 tested after 90 days.

at the interface area between coating and substrate. Over time, these cracks will reach the interface of coating and substrate and allow the penetration of moisture or impurities, which will deteriorate the properties of kaolin geopolymer-coated lumber woods. A sample with an S/L ratio of 0.9 (Figure 8.10c) clearly displayed an imprinted interfacial layer microimage that showed good adhesion between coating and subtrate. No open or closed pores were observed in samples tested after 90 days, which shows stabilty of the geopolymer gel formed in the early stages of this formulation. A sample with an S/L ratio of 1.0 (Figure 8.10d) exhibited fairly high physical properties due to the high amount of solid content that contributed to the sample bulk density. However, excessive unreacted particles lead to a poor and incomplete geopolymerization process. Uneven coating thickness is also observed after 90 days due to the formation of whitish efflorescence that is merely physically adhered to the coating layer. Micro-cracks are also observed on the surface of the sample, which would further contribute to the lowering of its bonding properties. The sample with the highest S/L ratio (Figure 8.10e) was also almost peel-off-ready when tested after 90 days.

It is clear through its microimages that the bonding properties of the sample with the unsuitable formulation change rapidly over time, proving the high solid content is unnecessary in geopolymer coating. With time, this led to formation of efflorescence that hindered further geopolymerization processes. Therefore, mechanical and adhesion strength dropped drastically with sample age. These optical data further support flexural properties that shows an S/L ratio of 0.9 is more suitable for kaolin geopolymer-coated lumber woods.

8.5.2 X-Ray Diffraction Analysis

Figure 8.11 presents the phase composition of fly ash geopolymer ceramic coatings at various temperatures. Before sintering, the phase shows a semi-crystalline phase, as indicated in the diffraction pattern between 20° and 40° (2θ). The content of quartz with the sharp peak indicates 26.721° (2θ) from the original materials. The intensity for unsintered coating is 566.9 cps. After sintering at 600°C, the XRD pattern shows a slight decrease in quartz intensity to 497 cps and a peak slightly shifting from 26.721° (2θ) for unsintered to 26.621° (2θ). The XRD pattern for temperature 1,500°C with the highest strength shows a different peak position at 34.81° (2θ). In order to quantify peak shifting, the difference in 2θ value of the sintering geopolymer ceramic coating was compared with the reference value of the unsintered coating. The quartz peak was decomposed after heat treatment at 1,500°C.

From the XRD pattern, it can be observed that the sharp peak of quartz in the geopolymer ceramic coating has distortion. During sintering, two kinds of processes occur: decomposition and phase transformations (Gonidanga et al., 2019). Due to the presence of iron (III) oxide compounds, the semi-crystalline phase's formation enhanced the flexural strength of the fly ash geopolymer ceramic coating.

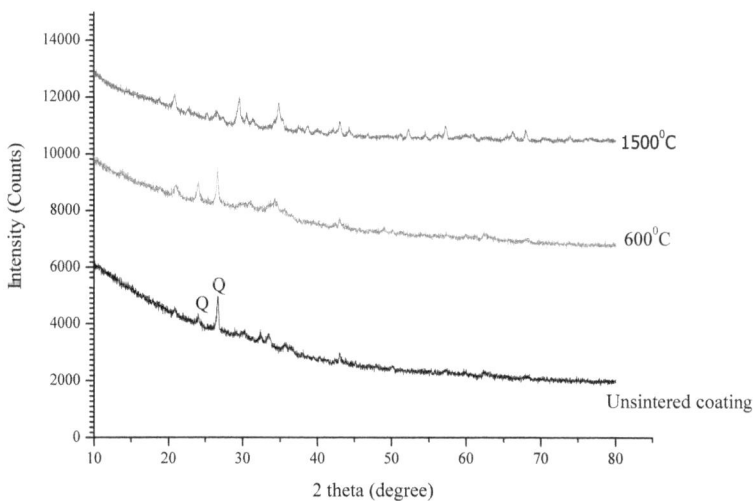

FIGURE 8.11 Phase composition of fly ash geopolymer ceramic coating at various temperatures (Q = quartz).

In the semi-crystalline phase, dislocations of atomic structure increase the mechanical properties of geopolymer ceramic coating.

Figure 8.12 shows the intensity changes of phase in kaolin geopolymers with an S/L ratio of 0.9 tested after 7, 28, and 90 days. As displayed in Figure 8.12, the presence of kaolinite peaks at 2θ values of 13.5°, 24.9°, 35.5°, 62.3°, and 68.3° was still present in kaolin geopolymers even after 90 days of testing. The typical representation of the amorphous phase of geopolymer was visible through the obvious broad hump at 2θ values in the range of 15°–45°. In addition, zeolites with peaks at 2θ values of 17.2°, 31.6°, 34.1°, and 52.3° were observed in all samples upon geopolymerization. Other existing peaks, such as anatase, sodalite, and quartz, also decreased by the day of testing from 7 to 90 days, as shown in Figure 8.12. Based on preliminary studies done earlier, the lowest intensity of remaining kaolinite peaks and a comparatively broader hump than other formulations suggested that kaolin geopolymers with an S/L ratio of 0.9 are the optimum conditions. The findings were in agreement with the mechanical strength analysis of kaolin geopolymers.

Varying the S/L ratio with a small range of difference does not majorly affect the phase changes in kaolin geopolymers, as not much difference was seen in the phase patterns of kaolin geopolymers tested after 7 days. Kaolinite peaks as shown in Figure 8.12, slightly dropped in terms of their intensity with increasing sample age. The broad hump features of the X-ray diffractogram also indicate the formation of aluminosilicate gel. The amorphous phase of kaolin geopolymers also contributes to their mechanical strength (Matalkah et al., 2020).

Kaolinite peaks, as shown in Figure 8.12, slightly dropped in terms of their intensity with increasing sample age, proving the prolonged geopolymerization process occurred over time. It is also predicted that this pattern will continue until the sample

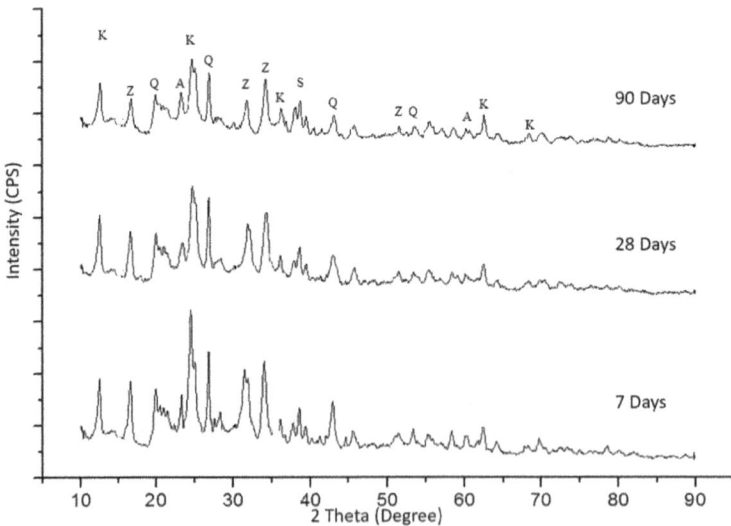

FIGURE 8.12   X-ray diffractogram of kaolin geopolymers with a solids-to-liquid (S/L) ratio of 0.90 tested after 7, 28, and 90 days.

age of 180 days, based on the findings of Heah et al. (2012). A noticeable reduction in zeolite peaks from 7 to 90 days of testing indicates the change of phase of kaolin geopolymers in terms of their crystallinity. Zeolite is known to represent a major part of the crystalline phase of kaolin geopolymers; thus, a decrease in zeolite indicates an increase in compressive strength over time. This proves that zeolite contributes to brittleness and increases the porous nature of kaolin geopolymers. Peaks such as anatase, sodalite, and quartz also decreased with increasing sample age days as a result of active dilution effect.

8.6 CONCLUSIONS

This research focuses on fly ash and kaolin as geopolymer source materials in coating applications. The best mix design of fly ash-based geopolymer paste was successfully tested on ceramic samples, while kaolin geopolymer paste was tested on lumber-grade wood substrates. The best design of geopolymer paste is determined by the ratio of fly ash/kaolin-to-alkaline activator (FA/AA) and the ratio of sodium silicate to sodium hydroxide (Na_2SiO_3/NaOH) solution. The best ratio is selected by testing the compressive strength of the geopolymer paste before applying it to the substrate sample. The condition of the coating material was tested at the temperature range of 600°C–1,600°C based on the high-temperature application purpose for fly ash geopolymer coating. The optimum mix design for fly ash geopolymer paste is fly ash to alkaline activator (FA/AA) and sodium silicate to sodium hydroxide (Na_2SiO_3/NaOH) ratio of geopolymer paste was obtained at 2.0 and 2.5 with a maximum compressive strength of 52 MPa. Kaolin geopolymer coating strength increased with increasing S/L ratios from 0.7 to 1.1, the highest compressive strength of 2.45 MPa achieved after 7 days was for a sample with a kaolin-to-alkaline activator ratio of 1.1, while a sample with an S/L ratio of 0.7 showed the lowest compressive strength of 0.45 MPa.

The potential of geopolymer as a coating material was supported by the fact that there is abundant industrial waste from power plants that is suitable to be used as a source of material for geopolymer and has excellent properties as a coating material.

REFERENCES

Ana María, A. G., & de Ruby Mejía, G. (2018). Eco-efficient repair and rehabilitation of concrete Infrastructures. In: *Assessment of corrosion protection methods for reinforced concrete, Woodhead Publishing Series in Civil and Structural Engineering* (pp. 315–353). United Kongdom: Woodhead Publishing.

Castillo, H., Collado, H., Droguett, T., Sánchez, S., Vesely, M., Garrido P., & Palma, S. (2021). Factors affecting the compressive strength of geopolymers: A review. *Minerals*, 11, 1317.

Davidovits, J. (1994). High-alkali cements for 21st century concretes. In: *Concrete Technology, Past, Present and Future*. American Concrete Institute, Farmington Hills, MI.

Fifinatasha Shahedan, N., Mustafa Al Bakri Abdullah, M., Mohd Ruzaidi Ghazali, C., Binhussain, M., Al Husaini, Kamarudin Hussin M., & Ramasamy, S. (2016). Morphology and properties of geopolymer coatings on glass fibre-reinforced epoxy (GRE) pipe. *MATEC Web of Conferences*, 7, 01069.

Gonidanga, B. S., Njoya, D., Lecomte-Nana, G., & Njopwouo, D. (2019). Phase transformation, technological properties and microstructure of fired products based on clay-dolomite mixtures. *Journal of Materials Science and Chemical Engineering*, 7, 11.

Haleh, R., Alireza, J., Soheil, J., Farhad, A., & Maryam, G. (2020), Rheology and workability of SCC. In: Rafar Siddique (ed.), *Self-Compacting Concrete* (pp. 31–63). Woodhead Publishing, Elsevier, United Kingdom.

Heah, C. Y., Hussin, K., Bakri, A. M., Binhussain, M., Luqman, M., Nizar, K. I., Ruzaidi, C. M., & Liew, Y. M. (2012). Study on solids-to-liquid and alkaline activator ratios on kaolin-based geopolymers. *Construction and Building Materials*, 35, 912–922.

Ikmal, H. A., Mohd Mustafa, A. A., Heah, C. Y., & Liew, Y. M. (2020). Behaviour changes of ground granulated blast furnace slag geopolymers at high temperature. *Advances in Cement Research*, 32(10), 465–475.

Klimaa, K. M., Schollbacha, K., Brouwers, H. J. H., & Qingliang, Y. (2022). Thermal and fire resistance of class F fly ash based geopolymers: A review. *Construction and Building Material Journal*, 323, 126529.

Lăzărescu, A., Mircea, C., Szilagyi H., & Baeră, C. (2019). Mechanical properties of alkali activated geopolymer paste using different Romanian fly ash sources. *MATEC Web of Conferences*, 289, 11001.

Matalkah, F., Aqel R., & Ababneh, A. (2020). Enhancement of the mechanical properties of kaolin geopolymer using sodium hydroxide and calcium oxide. *Procedia Manufacturing*, 44, 164–171.

Nikolov, A., Rostovsky, I., & Nugteren, H. W. (2017), Geopolymer materials based on natural zeolite. *Case Studies in Construction Materials*, 6, 198–205.

Papakonstantinou, C. G., & Balaguru, P. N. (2017). Geopolymer protective coatings for concrete. In: *International SAMPE Symposium and Exhibition (Proceedings)*, 52, 0891–0138 (USA).

Safiuddin, M., & Angelo Del, Z. (2017). Concrete damage in field conditions and protective sealer and coating systems. *Coatings*, 7(7), 90.

Wang, A., Fang, Y., Zhou, Y., Wang, C., Dong B., & Chen, C. (2022). Green protective geopolymer coatings: Interface characterization, modification and life-cycle analysis. *Materials*, 15(11), 3767.

Yong-Sing, N., Yun-Ming, L., Mustafa Al Bakri Abdullah, M., Hui-Teng, N., Hussin, K., Cheng Yong, H., Aida Mohd Mortar N., & Victor Sandu, A. (2020). Effect of solid-to-liquid ratio on thin fly ash geopolymer. *IOP Conference Series: Materials Science and Engineering*, 743, 012006.

Zhu, A. (2020). Feasibility study on novel fire-resistant coating materials, Master Theses, Missouri University of Science and Technology, Rolla, MO, USA.

9 Corrosion in Geopolymer Materials
A Case Study

Farah Farhana Zainal, Kamarudin Hussin,
Azmi Rahmat, and
Mohd Mustafa Al Bakri Abdullah
Universiti Malaysia Perlis

Ratna Ediati
Institut Teknologi Sepuluh Nopember

9.1 INTRODUCTION

The corrosion problem on the reinforcing bar in the concrete structure is the most crucial issue nowadays. It will cause surface deterioration on concrete structures and damage to the embedded steel reinforcing rods. This costs a lot of money to refurbish these structures. For instance, the average annual cost to rehab and repair these concrete structures damaged by corrosion for 15 years in the US is more than $6.3 billion (Tahershamsi, 2016). To overcome this problem, an environmentally friendly material called geopolymer is used. Geopolymer produces a three-dimensional aluminosilicate gel from inorganic alkali activation synthesis. This material has a few advantages, such as dimensional stability, fire resistance, durability, higher early strength, and a superior bond to reinforcement and aggregates (Nuruddin et al., 2016).

Nevertheless, less attention was given to corrosion behavior in geopolymer structural applications. In this research chapter, corrosion behavior at reinforcement bars embedded in geopolymer paste is studied. In this research study, two corrosion tests were performed to determine the corrosion performance of the reinforcement bar. The two corrosion tests conducted are Tafel extrapolation to determine the corrosion rate and open circuit potential (OCP) to determine the potential values. Phase analysis and microstructure analysis were also performed to prove the existence of the passive layer between the reinforcement bar and geopolymer paste.

9.2 RAW MATERIALS

The raw materials used in this study were fly ash, sodium hydroxide (NaOH) solution, sodium silicate (Na_2SiO_3) solution, and reinforcement bar (rebar). NaOH and Na_2SiO_3 solutions were acting as alkaline activators in this research.

DOI: 10.1201/9781003390190-9

9.2.1 Fly Ash

The fly ash used in this experiment is a low-calcium class F fly ash (ASTM C618, 2019) collected from the Sultan Azlan Shah Power Station, Manjung, Lumut, Perak, Malaysia. This fly ash has been used as a base ingredient in the production of geopolymer paste. Class F fly ash has a few advantages; for instance, it could lower concrete permeability and reduce chemical and sulfate attack, corrosion of reinforcement, and alkali silica reactions. This is due to fly ash class F, which is a high pozzolanic material and consists of almost 70% pozzolanic compounds.

9.2.2 Sodium Hydroxide (NaOH) Solution

The NaOH pellets used in this study were supplied by HmbG® Chemicals with a molar mass of 40.00 g/mol. It also has an absolute density of 2.13 g/cm^3, a melting point of 318.4°C, a boiling point of 1,390°C, and a purity of 97%–98%. It has a liquid moisture characteristic where the NaOH pellets will easily melt when exposed for a long time, even in a room temperature environment. NaOH solution was produced by the process of dissolving NaOH pellets with distilled water. The 12 M NaOH solution was used in this research. According to a study conducted by Kamarudin et al. (2011), Abdullah et al. (2013, 2021), Scrivener and Favier (2015), and Bakkali et al. (2016), the 12 M NaOH solution is better because it provides a high reading for compressive strength tests.

9.2.3 Sodium Silicate (Na$_2$SiO$_3$) Solution

Na$_2$SiO$_3$ used in this research is collected from South Pacific Chemicals Industries (SPCI), Malaysia. The chemical composition of Na$_2$SiO$_3$ is 9.40% Na$_2$O, 30.10% SiO$_2$, and 60.50% H$_2$O with a SiO$_2$/Na$_2$O modulus of 3.2, a specific gravity of 1.40 g/cm^3 at 20°C, and a viscosity of 0.40 Pa s at 20°C.

9.2.4 Reinforcement Bar (Rebar)

The reinforced concrete is designed based on the principle of reinforcement bars and concrete acting together across the force, where the concrete is strong in compression but weak in tensile strength. Therefore, reinforcement bars are needed to accommodate the tensile stresses imposed on them (Satyendra, 2015). The reinforcement bar used in this research is mild steel with a 12 mm diameter, as shown in Figure 9.1.

9.3 DESIGN OF EXPERIMENT

9.3.1 Alkaline Activator Preparation

The alkaline activator solution was produced 24 hours earlier for use with a mixture ratio of Na$_2$SiO$_3$/NaOH of 2.5. This decision was supported by past studies conducted by Castillo et al. (2021) and Kwek et al. (2021), which show that the ratio of 2.5 gives an excellent compressive strength value. Ratio is very important to get a homogeneous alkaline activator solution.

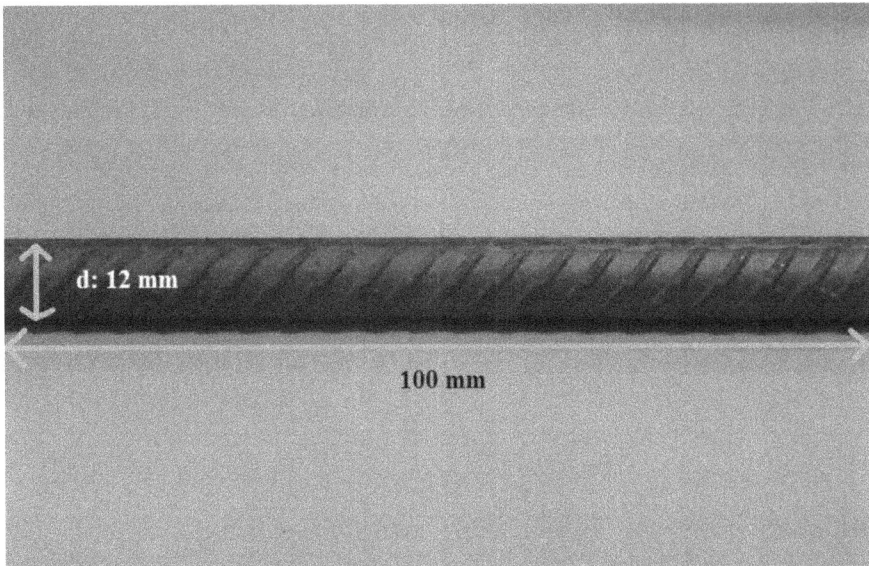

FIGURE 9.1   12 mm diameter reinforcement bar used in this study.

9.3.2 MIXING AND CURING PROCESS

An alkaline activator solution was blended with fly ash, and the mixture was stirred manually for 30 minutes. Then, the mixture was poured into 50 mm x 50 mm x 50 mm molds, except for the sample for the Tafel extrapolation test because the sample must be in liquid form. The mixture was poured in three stages into the mold and compacted 25 times at every stage by the rod. The reinforcement bars were placed in the middle of the samples. The samples were then left in the mold until they became harder. After the samples were hardened, the pH of the samples was determined using pH paper.

The samples were taken off the molds after 24 hours and placed in the oven for curing at 60°C for the next 24 hours. Curing is very important for the process of synthesis, especially its effect on compressive strength (Wan Mastura et al., 2012; Abdullah et al., 2014; Tennakoon et al., 2014; Nurruddin et al., 2018). Some previous studies also show that the best curing conditions were at 60°C for 24 hours, as it records the highest compressive strength compared to the other temperatures (Zharif Ahmad Jaafar, 2014; Abdullah et al., 2014).

9.4   CORROSION TESTING

A few corrosion tests were conducted to investigate the corrosion performance of reinforcement bars ingrained in geopolymer paste. The tests include Tafel extrapolation, OCP, phase analysis, and morphology analysis.

9.4.1 TAFEL EXTRAPOLATION ANALYSIS

This is one of the most frequent tests to find out the corrosion rate of a reinforcement bar. This test is carried out in conventional electrochemical cells comprising metals (exposed area of 1.0 cm²), reference electrode (saturated calomel reference electrode (SCE)) and geopolymer paste solution. Based on Figure 9.2, the graph was analyzed using the Tafel extrapolation test for geopolymer paste. The corrosion rate for the sample was 140.1×10^{-3} mpy. The E_{corr} value is -0.440 V SCE, while the I_{corr} value is 0.307 µA/cm². The βa and βc values, respectively, recorded $--1,600$ mV SCE and $--1,550$ mV SCE.

The I_{corr} value obtained through this study is interpreted according to the Table 9.1 analysis system. This analysis system has two conditions: construction site testing and laboratory testing. However, this study was conducted on a laboratory- scale, and the analytical system for laboratory tests was used. This geopolymer result of between 0.2- and 1.0 means that the damage is expected to occur in concrete or geopolymer in the next 10–15 years.

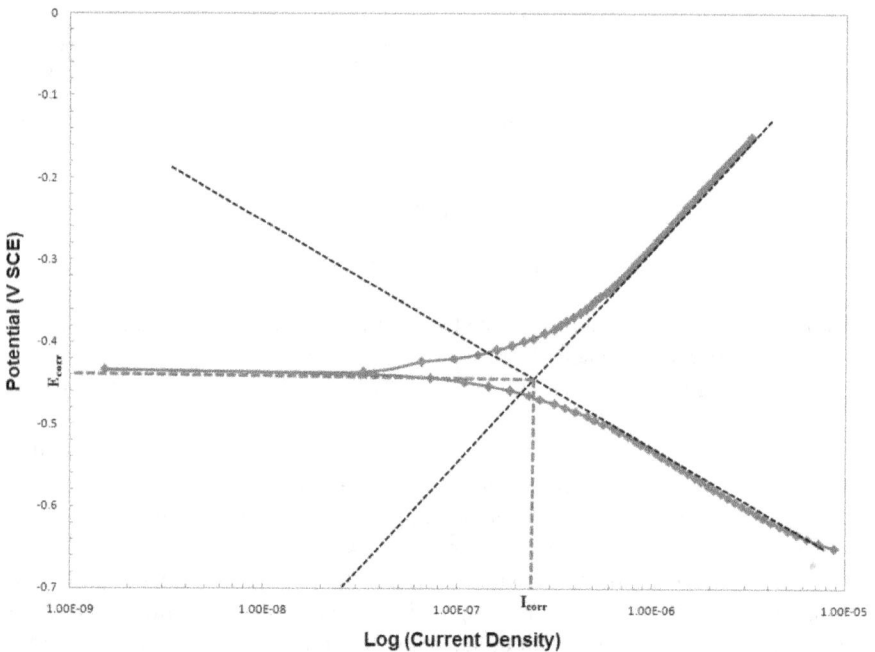

FIGURE 9.2   Tafel extrapolation graph for geopolymer paste.

TABLE 9.1

Interpretation of Guidelines for Icorr and Reinforcement Bar Steel Conditions (Bastidas et al., 2021)

I_{corr} (µA/cm²)	Referred for Lab Test Scale (Expected Level of Damage)
<0.2	No damage
0.2–1.0	Damage within 10–15 years
1.0–10	Damage within 2–10 years
>10	Damage within a year

9.4.2 Open Circuit Potential (OCP) Analysis

OCP is a technique that measures the potential value difference between working and reference electrodes. It is very important to determine the region of potential value in the Pourbaix diagram, whether it is in the passive, immunity, or corrosion regions. Figure 9.3 shows the maximum and minimum potential values for reinforced bar steel ingrained in geopolymer paste located in the passivity region. The geopolymer pH value has been examined using pH paper. The pH value for geopolymer paste in this experiment is 12, which means concrete should be in an alkaline environment. As reported by Davidovits, he declared that the pH of geopolymers shall be in the range of 11.5–12.5, depending on their formulation (Ralli et al., 2020). From Figure 9.3, the maximum potential value is 0.368 V, while the minimum potential value is 0.299 V.

9.5 PHASE ANALYSIS

It was conducted to prove the existence of a passive layer between the geopolymer paste and reinforcement bar. Phase analysis was done using X-ray diffraction (XRD) equipment. Figure 9.4 shows the XRD spectra of the carbon steel sample used as reinforcement bar steel in this research, embedded in the geopolymer sample. Based on the reinforcement bar steel spectrum, the ferric phase, Fe, shows the highest intensity at $2\theta = 45°$. In addition to the 45° angle, the ferric phase also exists at an angle of 65°. From the reinforcement bar spectrum embedded in the geopolymer sample,

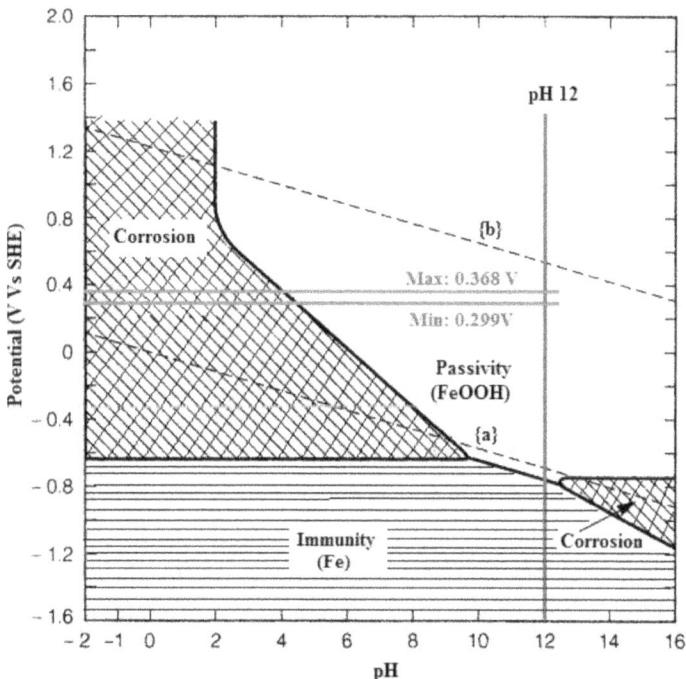

FIGURE 9.3 The Pourbaix diagram shows the maximum and minimum potential values for reinforced bar steel embedded in geopolymer paste.

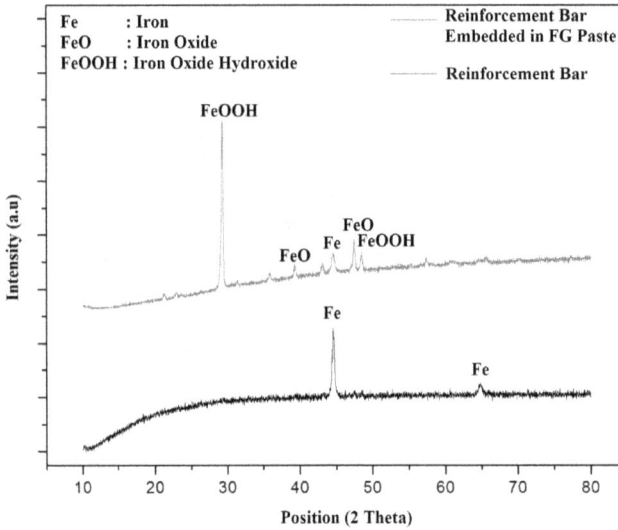

FIGURE 9.4 XRD pattern for reinforcement bars embedded and not embedded in geopolymer paste (Farhana Zainal et al., 2017).

phase analysis has also been performed to analyze the coating passively formed from the reaction between steel, oxygen, and geopolymer. Figure 9.4 also shows the iron oxyhydroxides (FeOOH) found in a reinforcement bar that is embedded in geopolymer paste.

The deoxydylation mechanism under any circumstances, whether in the form of a solution or dry final product for FeOOH, is hematite (Fe_2O_3) (Song et al., 2016). Dehydroxylation means the loss of water molecules in the hydroxyl ion structure during the heating process. However, the phase analysis shows FeOOH was found on the reinforcement bar surface. This is because the experimental samples are only 7 days old, while the dehydroxylation process takes a long time to change to the Fe_2O_3 end product. The higher the intensity of a sample, the better the crystallization. The intensity change also occurs at an angle of 45°, where the tip of the iron is lower than the original because a chemical reaction has occurred between the steel and the geopolymer. Ferum on the reinforcement bar changes to FeOOH. Therefore, the analysis shows FeOOH is high while iron is very low in the sample.

9.6 MICROSTRUCTURE ANALYSIS

Microstructure in this research was carried out by using a scanning electron microscope (SEM) to prove the existence of the passive layer between the reinforcement bar and geopolymer paste. The presence of a passive layer on the reinforcement bar can also be demonstrated by the SEM micrograph at 700X magnification, as shown in Figure 9.5. This figure shows a passive layer formed between reinforcement bars, and geopolymer paste with a thickness of 84.5 μm. The layer will act as a protector against reinforcement bars from other corrosion agents.

FIGURE 9.5 SEM micrographs of passive layers formed on reinforced bar steel embedded in geopolymer paste at 700X magnification.

9.7 CONCLUSIONS

It can be concluded that geopolymerization is a field that is getting more and more attention. This is because the geopolymer itself has many advantages, especially in curbing the occurrence of corrosion on concrete structures, especially in areas exposed to the seawater environment. The passive layer formed from the reaction between pozzolonic materials and reinforcing steel bars could protect the reinforcing steel bars from aggressive agents that can cause corrosion. It is because the reinforcement steel bar in geopolymer concrete was coated with a strong and adherent silicate membrane. The function of a silicate membrane is to protect the reinforcement bar from corrosion, although in neutral solutions.

REFERENCES

Abdullah, A., Hussin, K., Abdullah, M. M. A. B., Yahya, Z., Sochacki, W., Razak, R. A., Błoch, K., & Fansuri, H. (2021). The effects of various concentrations of naoh on the inter-particle gelation of a fly ash geopolymer aggregate. Materials, 14, 1111.

Abdullah, M. M. A. B., Razak, R. A., Yahya, Z., Hussin, K., Ming, L. Y., & Ahmad, H. C. Y. M. Z. (2013). *ASAS Geopolimer.* Teori & Amali, Penerbit Universiti Malaysia Perlis (UniMAP), Perlis.

Abdullah, M. M. A. B., Yahya, Z., Abdullah, A., Razak, R. A., Hussin, K., & dan Muhammad Faheem Mohd Tahir, L. J. (2014). *Bahan Geopolimer: Pemprosesan, Pencirian dan Aplikasi.* Penerbit Universiti Malaysia Perlis (UniMAP), Perlis.

American Society for Testing and Materials (ASTM). (2019). *Designation: C618: Standard Specification for Coal Fly Ash and Raw or Calcined Natural Pozzolan for Use in Concrete*, ASTM International, West Conshohocken, PA.

Bakkali, H., Ammari, M., & Frar, I. (2016). NaOH alkali-activated class F fly ash: NaOH molarity, curing conditions and mass ratio effect. *Journal of Materials and Environmental Science*, 7(2), 397–401.

Bastidas, D. M., Martin, U., Bastidas, J. M., & Ress, J. (2021). Corrosion inhibition mechanism of steel reinforcements in mortar using soluble phosphates: A critical review. *Materials,* 14. doi:10.3390/ma14206168

Castillo, H., Collado, H., Droguett, T., Sánchez, S., Vesely, M., Garrido, P., & Palma, S. (2021). Factors affecting the compressive strength of geopolymers: A review. *Minerals*, 11, 1317.

Davidovits, J. (2005). Geopolymer chemistry and sustainable development. The poly (sial-ate) terminology: A very useful and simple model for the promotion and understanding of green-chemistry. In: Joseph Davidovits (ed.), *Proceedings of World Congress Geopolymer* (235 p.). Geopolymer Institute.

Farhana Zainal, F., Amli, S. F. M., Hussin, K., Rahmat, A., & Abdullah, M. M. A. B. (2017). Corrosion studies of fly ash and fly ash-slag based geopolymer. *IOP Conference Series: Materials Science and Engineering*, 209, 012026.

Kamarudin, H., Mustafa Al Bakri, A. M., Binhussain, M., Ruzaidi, C.M, Luqman, M., Heah, C. Y., & Liew, Y. M. (2011). *Preliminary Study on Effect of NaOH Concentration on Early Age Compressive Strength of Kaolin-Based Green Cement, International Conference on Chemistry and Chemical Process* (Vol. 10). IACSIT Press, Singapore.

Kwek, S. Y., Awang, H., & Cheah, C. B. (2021). Influence of liquid-to-solid and alkaline activator (sodium silicate to sodium hydroxide) ratios on fresh and hardened properties of alkali-activated palm oil fuel ash geopolymer. *Materials (Basel)*, 14(15), 4253.

Nurruddin, M. F., Haruna, S., Mohammed, B. S., & Galal Shaaban, I. (2018). Methods of curing geopolymer concrete: A review. *International Journal of Advanced and Applied Sciences*, 5(1), 31–36.

Nuruddin, M. F, Malkawi, A. B., Fauzi, A., Mohammed, B. S, & Almattarneh, H. M (2016). Geopolymer concrete for structural use: Recent findings and limitations. *IOP Conference Series: Materials Science and Engineering*, 133, 012021.

Ralli, Z., & Pantazopoulou, S. (2020). State of the art on geopolymer concrete. *International Journal of Structural Integrity*, 12(4), 511–533.

Satyendra, November 14, 2015. Concrete and reinforced concrete. *IspatGuru*. https://www.ispatguru.com/concrete-and-reinforced-concrete/

Scrivener, K., & Favier, A. (2015). *Calcined Clays for Sustainable Concrete: Proceedings of the 1st International Conference on Calcined Clays for Sustainable Concrete*. Springer, Dordrecht.

Song, X., & Boily, J.-F. (2016). Surface and bulk thermal dehydroxylation of FeOOH polymorphs. *The Journal of Physical Chemistry A*, 120(31), 6249–6257.

Tahershamsi, M. (2016). Structural Effects of Reinforcement Corrosion in Concrete Structures. Thesis for the Degree of Doctor of Philosophy, Department of Civil and Environmental Engineering Division of Structural Engineering Concrete Structures, Chalmers University of Technology, Gothenburg, Sweden.

Tennakoon, C., Shayan, A., Sagoe-Crentsil, K., & Sanjayan, J. G. (2014). Importance of reactive SiO_2, AlO_3 and Na_2O in geopolymer formation. In: *Austroads Bridge Conference*, Sydney, NSW, Australia.

Wan Mastura, W. I., Kamarudin, H., Nizar, K., Abdullah, M. M. A. B., & Mohammed, H. (2012). The effect of curing time on the properties of fly ash-based geopolymer bricks. *Advanced Materials Research*, 626, 937–941. 10.4028/www.scientific.net/AMR.626.937.

Zharif Ahmad Jaafar, A. (2014). *Mechanical Behavior of Fly Ash Based Geopolymer Cement as Well Cement*. Universiti Teknologi Petronas, Perak.

10 Geopolymer Soil Stabilization
A Case Study

Liyana Ahmad Sofri,
Muhammad Faheem Mohd Tahir, and
Mohd Mustafa Al Bakri Abdullah
Universiti Malaysia Perlis

Thanongsak Imjai
Walailak University

I Nyoman Arya Thanaya
Udayana University

10.1 INTRODUCTION

There are a few types of soil stabilization that refer to various methods that can improve the mechanical and physical characteristics of soil for a specific purpose, including construction or other purposes. This includes the process of modifying the texture and plasticity of the soil by adding and removing sand, clay, and other materials. It may also involve compacting the soil to enhance its load resistance. Alternatively, inorganic binders (such as cement) may be used to make problematic soil more workable or to achieve better strength and durability. This chapter will discuss the benefits and drawbacks of soil stabilization, the types of materials that can be utilized for this purpose, and the application of geopolymer in soil stabilization.

A controlled modification of the texture, structure, and physico-mechanical properties of the soil is one term of the method named soil stabilization. It is relatively uncommon for natural soil to be perfectly suitable for construction materials in their "as raised" condition without any soil assessment modification. In addition to this, the majority of soil modification involves some type of compaction to form a material that is strong and stable. Stabilization techniques can be roughly categorized as physical, mechanical, and binding methods (Behnood, 2018). Therefore, almost all construction materials are stabilized in some form; nonetheless, the term "stabilization" is generally reserved for the application of inorganic binder additions exclusively. Physical stabilization is the addition or subtraction of different soil fractions to modify the particle size distribution and plasticity of the soil to improve its physical properties.

Mechanical stabilization is the technique of enhancing the soil's characteristics by modifying its gradation. This method involves the compaction and densification of soil through the application of mechanical energy via rollers, rammers, vibration methods, and often blasting. This process is based on the specific qualities of the soil material for soil stability. Two or more types of natural soils are combined to produce a composite material that is superior to the sum of its parts. Mechanical stabilization is formed by adding or combining soils of two or more gradations to create a material that meets the specific requirements.

Stabilization can enhance the shear strength of a soil and/or manage its shrink-swell capabilities, thereby increasing the soil subgrade's ability to withstand pavements and foundations. Saturated soils have higher plasticity, and it is widely known that soil strength and bearing capacity are substantially higher in the dry state than in the saturated condition. The most frequent state that may be observed is one of partial saturation; nevertheless, the specific amount of water content can vary greatly depending on the current weather conditions or groundwater level.

Portland cement, non-hydraulic lime, hydrophobic admixtures, and occasionally bituminous emulsion are the types of soil stabilization that are utilized the most frequently. According to the Australian Earth Building Handbook HB195 (Meek and Elchalakani, 2021), a non-hydraulic lime should be used to stabilize cohesive soils, and hydraulics such as Portland cement and bituminous stabilizers should be used for granular soils. This is one general approach to the process of soil stabilization. Figure 10.1 is a graph that is used to sum up the criteria for various stabilization procedures. The plasticity index and the relative amount of cohesive material (percent of soil mass below 80 mm particle diameter) are indeed crucial aspects. Therefore, this graph can be used to illustrate the selection criteria.

Aside from lime and cement, there are various other alternatives for soil stabilization, such as fiber reinforcement, polymers, and geopolymers. The following subsections will discuss these in details for its composition, properties, and binding mechanisms.

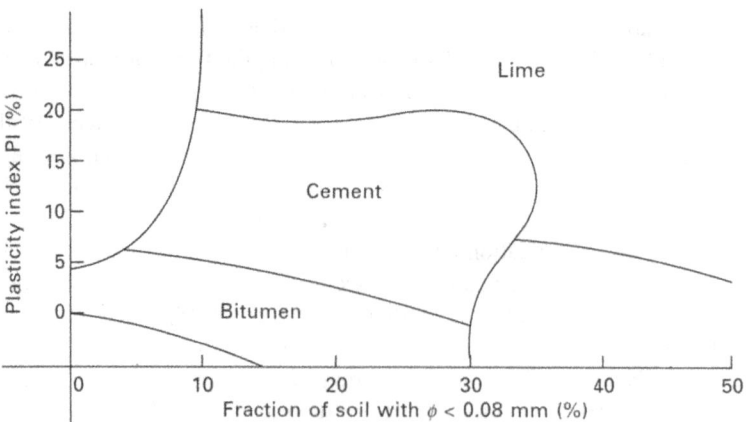

FIGURE 10.1 Selection criteria for common stabilizers with reference to soil characteristics.

10.2 ADVANTAGES AND DISADVANTAGES OF SOIL STABILIZATION

When it comes to construction, new investment projects frequently encounter undesirable soil conditions. It is essential to stabilize the underlying soil so that it can be incorporated into newly constructed civil engineering structures. New soil improvement methods and geotechnical engineering materials have enabled several methods for improving weak soil substrates. The improvement of the geotechnical properties of soils should offer the opportunity for upcoming construction projects to make effective and economical use of those soils as foundations for their construction (Brasse and Tracz, 2018).

Stabilized soils can have a higher status in some areas, such as developing countries, and can be distinguished from "conventional" earth materials. They are frequently used in "modern" structures. The following benefits and drawbacks should be properly examined before deciding whether to stabilize the soil mixture.

10.2.1 ADVANTAGES

- Significantly improves both the durability and strength of the soil, especially in areas where the existing soil is weak. According to Ghadir and Ranjbar (2018), the characteristics of the treated soil have an effect on the increase in strength over time. Soil pH, natural drainage, degree of weathering, organic carbon content, excessive amounts of exchangeable sodium, clay mineralogy, silica-sesquioxide ratio, presence of carbonates, extractable iron, and silica-alumina ratio are some of these properties.
- It can reduce or eliminate the need for expensive surface treatment or rendering.
- Plasticity index has decreased. The change in soil nature (granular nature after flocculation and agglomeration) is related to the reduction in plasticity, and the modified soil is as crumbly as silt soil. This is why it has a low surface area and a low liquid limit due to the plasticity of the lime (Nnochiri et al., 2018).

10.2.2 DISADVANTAGES

- The cost of raw materials has risen; the soil is free or low-cost, while cement is relatively pricey.
- The stabilization materials needed may not be readily available in some developing countries or may be expensive to transport.
- Mixing and constructing are two operations that might become more complicated depending on the type of stabilizer that is used. This can increase the probability that difficulties will emerge, which can have consequences for both the schedule and the budget.
- The use of cement and lime, both of which have the potential to have adverse effects on the environment, can lead to an increase in the embodied energy of the wall materials as well as the CO_2 emissions.
- Lime and cement are both hazardous substances that can cause skin and eye burns, which is not good for health and safety.

10.3 CONVENTIONAL SOIL STABILIZATION

10.3.1 LIME STABILIZATION

The oxides and hydroxides of calcium and magnesium are commonly referred to as lime. Commercial production of high calcium lime involves calcining carbonate rock minerals in the form of limestone or crushed chalk. It can also be manufactured commercially as dolomitic lime (Ca $(OH)_2$ + Mg $(OH)_2$), which is composed of calcium and magnesium oxides that are then pressure hydrated. Calcination of high calcium carbonate rock minerals occurs above 900°C and atmospheric pressure, forming highly reactive calcium oxide (CaO), which can subsequently be hydrated to the hydroxide form Ca $(OH)_2$. This is accomplished by either utilizing steam to create a dry hydrate or slaking in water to create a wet putty. Figure 10.2 represents the calcination–hydration–carbonation cycle, also known as the "lime cycle." Hydrated limes are commonly used for soil stabilization on temporary road surfaces and subbases, especially when the subsoil contains a high proportion of cohesive fine aggregate or clay. Civil engineers in the United States developed this road-building technology, first used by the Romans and other ancient civilizations, in the 1920s.

Utilizing lime can greatly enhance engineering properties. There are two types of improvement: modification and stabilization. To some extent, lime can be used to modify practically all fine-grained soils, but the most significant improvement occurs in clay soils with moderate to high plasticity. The exchange of calcium cations supplied by the hydrated lime for the typically existing cations adsorbed on the surface of the clay mineral is the main cause of modification. The clay surface mineralogy is modified as it reacts with the calcium ions to generate cementitious products when the hydrated lime reacts with the clay mineral surface in a high pH environment. As a result, plasticity and swelling are decreased, moisture-holding capacity is reduced, and stability is improved.

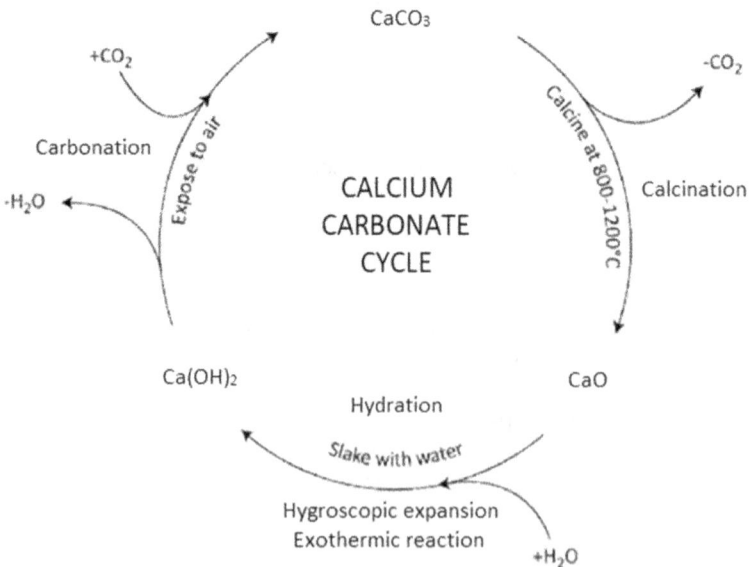

FIGURE 10.2 The calcination–hydration–carbonation cycle of calcium carbonate.

FIGURE 10.3 Limestone.

Stabilization occurs when the correct amount of lime is added to reactive soil. Stabilization differs from modification in that a significant increase in long-term strength is obtained through a pozzolanic reaction. As the calcium from the lime reacts with the aluminates and silicates solubilized from the clay mineral surface, calcium silicate hydrates and calcium aluminates are formed. This reaction can occur immediately and is critical for some of the effects of the modifications. However, the complete pozzolanic reaction might last for a very long time, even for several years. As a result, the addition of lime to certain soils might result in an enhanced high-strength gain. The key to pozzolanic reactivity and stability is soil, which is reactive and a great technique for mix design. Stabilization can contribute to very significant increases in resilient modulus values, improvements in shear strength, an increase in strength over time, and long-term durability even after decades of use. Figure 10.3 shows the limestone that is commonly used in soil stabilization.

10.3.2 CEMENT AND POZZOLANS

Portland cement was named after the Isle of Portland in Dorset, England, where it was first produced in the early 19th century. It is used to make concrete, mortar, and grout for a wide variety of construction projects. The chemical composition and physical properties of Portland cement can vary depending on the specific type and brand, but it typically has a high compressive strength and is resistant to water and chemicals. Portland cement is a type of hydraulic binder that is commonly used in the construction industry. It is produced by heating a mixture of limestone, clay, and other minerals to a high temperature, which triggers chemical reactions that convert the raw materials into a new substance called clinker. The clinker is then ground into a fine powder, which is the Portland cement that has been used (Angriawan, 2019). When Portland cement is mixed with water, it forms a paste that hardens and sets through a process called hydration. This process involves chemical reactions between the cement and the water, which create new compounds that bind the particles together and form a solid structure. Once the cement has hardened, it retains its strength and stability even when submerged in water, making it a popular choice for use in construction projects that require strong, durable materials.

The raw ingredients of the cement are calcium carbonate-bearing rock clays and iron ore (hematite). These materials are pulverized into fine powder and then heated in a rotary kiln at temperatures ranging from 700°C to 1,400°C. The final product, known as clinker, also contains four primary reactive compounds: tricalcium silicate (C3S), dicalcium silicate (C2S), tricalcium aluminate (C3A), and tetracalcium aluminoferrite (C4AF), as shown in Table 10.1. Of these four compounds, C3A is the most reactive and can cause the cement to set too quickly if not properly controlled. To prevent this, a small amount of gypsum ($CaSO_4 \cdot 2H_2O$) is added to the clinker before grinding it into a fine powder. The gypsum reacts with C3A to form insoluble calcium sulfoaluminate, which slows down the setting time and allows for more workable and consistent cement.

The quantity of each amount of clinker can be determined by the raw materials that were used and how long they were heated at certain temperatures during the production process. The cement that is made can then be used for building construction, walls, and pavements, among other things.

The composition of the clinker can be used to determine the properties of the resulting cement. For example, a greater C3S content can raise the setting speed and percentage of strength gain of the cement, while a low C3A content can result in sulfate-resistant cement. Similarly, a higher C2S component can produce cement with a lower heat of hydration, which is important in reducing the risk of thermal cracking in large concrete structures.

Some of the clay particles coating the coarse aggregates can become detached and dispersed into the aqueous phase when Portland cement and water are added to the soil. This can have an impact on the exothermic hydration of cement paste, thereby influencing the setting time and strength gain of the mixture. The remaining clay particles that adhere to aggregate surfaces can also affect the interfacial transition zone (ITZ) at the interface between cement paste and aggregate surfaces. This zone plays an important role in the mechanical performance of the resulting mixture, as it is the area where stresses are transferred between the cement paste and the aggregate particles.

The proportional ratio of clay particles that become dispersed within the aqueous solution and become part of the hardened cement paste microstructure is dependent on several factors, including the mineralogy and particle size distribution of clays.

TABLE 10.1
Reactive Compound in Clinker

Compound	Chemical Equation
Tricalcium silicate	$3CaO \cdot SiO_2$
Dicalcium silicate	$2CaO \cdot SiO_2$
Tricalcium aluminate	$3CaO \cdot Al_2O_3$
Tetracalciumaluminoferrite	$4CaO \cdot Al_2O_3 \cdot Fe_2O_3$

FIGURE 10.4   Cement powder.

The presence of certain types of clays, such as kaolinite, can lead to an increase in the amount of clay dispersed into the aqueous solution and a corresponding increase in the strength of the resulting mixture. Figure 10.4 presents the cement powder that is commonly used in mix design for soil stabilization.

10.3.3   BITUMINOUS BINDERS

Bitumen is a highly viscous black or dark brown material that is a by-product of oil refining. It is composed of long-chain complex hydrocarbons, with a typical carbon content of 82%–88% and a hydrogen content of 8%–11%. The remaining components are typically made up of small amounts of sulfur, oxygen, and nitrogen. The chemical composition of bitumen is comprised of four fractional components (Wang et al., 2021):

- *Asphaltenes*—These are the heaviest and most complex fraction of bitumen, consisting of high molecular weight hydrocarbons and heteroatoms such as sulfur, nitrogen, and oxygen. Asphaltenes are responsible for the high viscosity and adhesion properties of bitumen.
- *Resins*—These are intermediate weight molecules that are composed of aromatic and aliphatic hydrocarbons. Resins help to provide bitumen with its elasticity and adhesion properties.
- *Aromatics*—These are lighter weight molecules that are composed of aromatic hydrocarbons. Aromatics contribute to the solvency and fluidity of bitumen.
- *Oils*—These are the lightest fraction of bitumen, consisting of low molecular weight hydrocarbons. Oils provide bitumen with its fluidity and low-temperature flexibility.

When bitumen is used to stabilize granular materials, the resulting material is called a bitumen stabilized material (BSM). BSMs exhibit similar behavior to unbound granular materials, but with much increased cohesive strength and decreased sensitivity to moisture. In BSMs, the bitumen disperses only among the finest particles, resulting in a bitumen-rich mortar between the coarse particles, while the bigger aggregate particles are not coated with bitumen. The advantages of bituminous emulsion for soil stabilization are as follows (Asphalt Academy, 2009):

- Improved strength and stability: Bituminous emulsions serve as a binding agent and enhance the strength and stability of soil properties. The emulsion creates a more compact and stable substance by filling the spaces between the soil particles.
- Reduced moisture sensitivity: The bituminous emulsion reduces the sensitivity of soil materials to moisture by preventing water from penetrating the soil. This can also avoid damage from freeze-thaw cycles and boost the overall durability of the materials.
- Reduced dust generation: Bituminous emulsions can be utilized to minimize dust emission from other soil surfaces, such as unpaved roadways. The amount of loose dust produced is decreased by the emulsion's ability to bind the soil particles together.
- Cost-effective: Bituminous emulsion can be a cost-effective way to stabilize the soil because it can often be used with the local soil and does not require bringing in any extra materials.
- Environmentally friendly: Bituminous emulsion can be used to stabilize soil in a way that is good for the environment because it can reduce the need for new materials and help reduce the carbon footprint of construction projects.

The mechanisms involved in the stabilization of soil with a bituminous material are significantly different from those involved with cement or lime. The primary purpose of bitumen is to add stability to the soil. The unconfined compressive strength (UCS) test is typically used to evaluate the strength of soils that have been stabilized with bituminous materials. This test measures the maximum axial stress that can be applied to a soil sample before it fails under unconfined conditions. The triaxial test, which measures the shear strength of the soil, is not typically used to evaluate the strength of bituminously stabilized soils.

10.3.4 Synthetic Binders, Polymers and Adhesives

These synthetic binders and organic chemicals can improve the properties of soil in a number of ways. Gypsum can increase the strength of soil and improve its workability. Tetrasodium pyrophosphate can increase soil density and reduce plasticity. Calcium, sodium, and potassium salts can improve soil structure and stability. Fatty polyamides, lignin, and casein can increase water resistance and reduce erosion (Huang et al., 2021). Polymers can provide a range of benefits depending on their specific properties, such as improving strength, reducing water absorption, and enhancing durability.

Polymers can be mixed with soil as a liquid to fill the pores of the soil and strengthen its structure. To utilize polymeric stabilization, the following conditions must be met:

- Able to properly bond with soil particles while penetrating the soil matrix.
- Able to cure and form a stable structure within the soil matrix.
- Must be able to withstand environmental factors such as temperature changes, moisture content, and chemical reactions.
- Must be able to resist mechanical stresses and maintain its structural integrity under load.
- Cost-effective compared to other soil stabilization methods.
- Compatible with the soil type and any other additives used in the stabilization process.

10.3.5 Fiber Reinforcement

After the Second World War, there was a greater focus on soil stabilization in order to support overseas operations based on the rapid hardening of soft soils to facilitate military traffic. Even though cement and lime remained the most popular stabilizing agents, research into non-traditional stabilizing techniques, such as fiber reinforcing, which can be used alone or in combination with conventional stabilizing admixtures, was initiated (Dhar and Hussain, 2019). Often at concentrations of up to 1 wt.% of soil, earth reinforcement techniques with fibers are utilized to increase the shear strength and stability of soils. These composite soil-fiber materials can be used for retaining walls, embankments, roads, and other construction projects because of their comparatively high tensile strength.

Typically, soil-fiber reinforcement is accomplished by combining the fibers to create a random orientation within the soil structure. This technique has the following benefits: (i) different materials can be used as soil-fiber reinforcement; (ii) minimal additional mixing equipment is required; (iii) recycled, waste, and by-product fibers can be used; (iv) it is compatible with a variety of soil types; and (v) large soil volumes can be treated.

The main drawbacks of this technology are that it can only be used at shallow depths and that some fibers degrade over time due to biodegradation (Prambauer et al., 2019). Jute, sisal, bamboo, lumber, cotton, glass, wool, and shredded rubber fiber are just a few of the various natural and synthetic fiber kinds that can be used to stabilize soil.

10.4 APPLICATION OF GEOPOLYMER IN SOIL STABILIZATION

On many civil engineering construction sites, highly compressible or soft soils are encountered, which lack the essential strength to support loads throughout construction or service life (Sorsa et al., 2020). Geopolymerization is a chemical stabilization process that can be used to improve the strength and stiffness of soft soils. Geopolymerization involves mixing a source material (such as fly ash, slag, or other industrial waste) with an alkaline solution (usually a solution of sodium or potassium hydroxide and sodium or potassium silicate) to create a geopolymer binder. This binder can then be mixed with the soft soil to create a stabilized mixture. One of the techniques to stabilize the soil is by geopolymerization process.

10.4.1 MECHANISM

In the late 1970s, Joseph Davidovits is credited with creating the term "geopolymer" to characterize a type of amorphous, three-dimensional alumina-silicate binder materials generated by the reaction of aluminosilicate powders and alkaline solutions. These materials are similar to traditional cementitious materials in that they can harden and bind together other materials, but they differ in their chemical composition and the way they are formed (Davidovits, 1991). Geopolymers are produced through geopolymerization, which is the interaction of aluminosilicate minerals with an alkaline solution to form a three-dimensional, cross-linked network of mineral-like molecules. The resulting material has properties that are similar to traditional cementitious materials, such as high compressive strength, durability, and resistance to chemical attack, but with the added benefits of being environmentally friendly and sustainable (Hamzah et al. 2015).

Geopolymerization can be broken down into two main steps that occur sequentially during the reaction. The initial step involves the dissolution of amorphous alumina-silicate materials in an alkaline solution, such as sodium or potassium hydroxide or silicate, to form reactive silica and alumina species. This step is often referred to as the "activation" stage, as it creates the reactive components that will later form the geopolymer binder. The next step involves the polycondensation of the reactive species into amorphous or semi-crystalline oligomers, which are intermediate compounds that are formed as the geopolymerization reaction progresses. These oligomers then continue to polymerize and harden into the final synthetic alumina-silicate materials, which form the geopolymer binder (Yuan et al., 2021).

10.4.2 PROCESS AND IMPLEMENTATION

X-ray fluorescence (XRF). This approach is used to identify the elemental composition of a sample using a non-destructive method. The elemental composition of the specimen can then be determined by analyzing these X-rays.

X-ray diffraction (XRD). It is a technique for analyzing the crystal structure of a sample. XRD works by directing a beam of X-rays at the sample and measuring the pattern of X-ray diffraction that is produced.

Scanning electron microscope (SEM). It operates by scanning a focused beam of electrons across the surface of the sample and measuring the electrons released by the sample.

Liquid Limit Test. The liquid limit test is a laboratory procedure that measures the plasticity of soil. The plasticity of soil is a measure of its ability to deform without cracking or breaking. The test involves taking a sample of the soil and placing it in a special cup called a liquid limit device. This number is called the liquid limit, and it represents the moisture content at which the soil changes from a plastic state to a liquid state. The liquid limit test is important in geotechnical engineering because it can provide information about the behavior of soils under different loading conditions. Several organizations have standardized the liquid limit test, including the American Society for Testing and Materials (ASTM) and the International Society for Soil Mechanics and Geotechnical Engineering (ISSMGE). These organizations provide detailed procedures for conducting the test and interpreting the results, which can be used to ensure consistency and accuracy in geotechnical testing.

Plastic Limit Test. The plastic limit test is a laboratory procedure for determining the plastic limit of soil. At the plastic limit, soil transitions from a plastic to a semi-solid condition. It is an important parameter for soil classification and engineering design. The plastic limit is an important parameter for soil classification and engineering design. The plastic limit can also be used to calculate the plasticity index, which is a measure of the range of water contents over which the soil can exist in a plastic state. Several organizations have standardized the plastic limit test, including the American Society for Testing and Materials (ASTM) and the International Society for Soil Mechanics and Geotechnical Engineering (ISSMGE) (Syed Zuber et al., 2015).

10.4.3 PROPERTIES AND PERFORMANCE

XRF Properties. The main constituents of clays are shown in Table 10.2. All samples contain significant amounts of alumina (Al_2O_3) and silica (SiO_2). The major components for Clays 1, 2, and 3 are 93.5%, 75.7%, and 84.6% of the total contents, respectively. The main criteria for the geopolymerization process are materials rich in Si and Al (Cevik et al., 2018). Chindaprasirt et al. (2007) studied course lignite with high calcium fly ash used as geopolymer material in geopolymer mortar, where the total percentage of Al_2O_3 and SiO_2 was 59.5%, and the result showed an increase in geopolymer mortar strength. According to research by Elimbi et al. (2011), the use of varied kaolinite clay amounts increased the strength of geopolymer cement, with total Al_2O_3 and SiO_2 ranging from 72% to 85%, respectively. Ismail et al. (2014) found that the synthesis of geopolymer phases employing blast furnace slag and fly ash will result in values of 47.48% Al2O3 and 87.84% SiO2, respectively.

XRD Observation. Figure 10.5a–c shows the X-ray diffractogram of the soils. According to Figure 10.5a, the mineralogical component in Soil 1 is kaolinite ($Al_2Si_2O_5(OH)_4$). Low quartz (SiO_2) is the mineralogical component in Figure 10.5b and c, which, respectively, represented Soil 2 and Soil 3. These findings support the elemental composition of clays from earlier. According to research done by Elimbi et al. (2011), the kaolinite clays used in the production of geopolymer cement were the basic clays and contained quartz. According to research by Ismail et al. (2014) on the XRD pattern of unreacted fly ash, quartz was the primary constituent. It has been demonstrated that quartz can contribute a significant amount of silicon

---

TABLE 10.2
Major Element Composition of Clay Samples

Major Oxides	Al_2O_3	SiO_2	K_2O	CaO	TiO_2	Fe_2O_3	Si/Al Ratio
Clay 1 (wt.%)	37.60	55.90	2.66	–	0.84	2.40	1.50
Clay 2 (wt.%)	21.00	54.70	3.62	4.04	1.40	13.84	2.60
Clay 3 (wt.%)	21.10	63.50	2.31	0.90	1.80	9.50	3.00

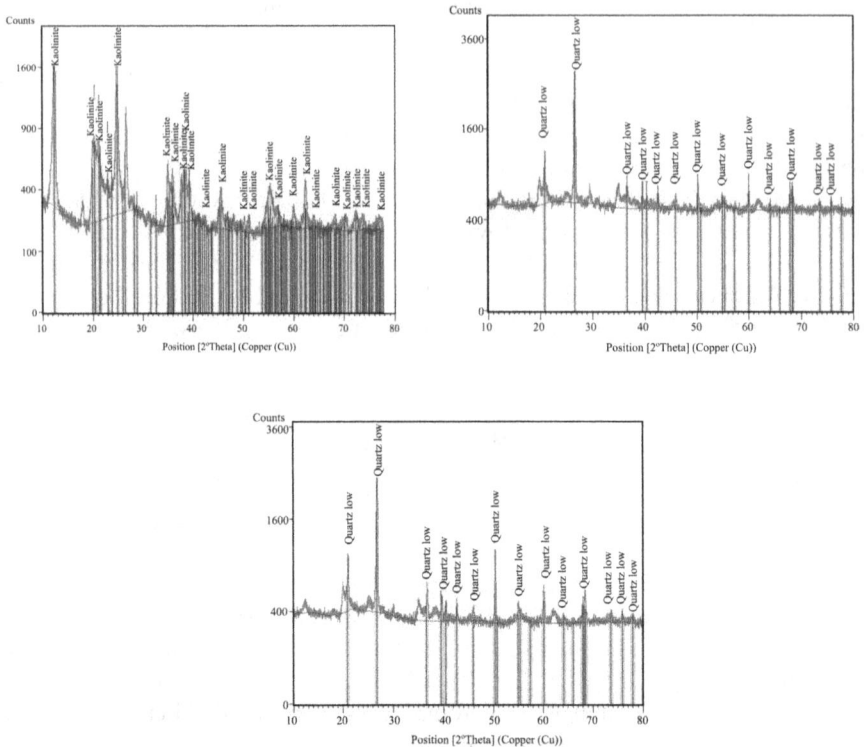

FIGURE 10.5   XRD pattern of analyzed samples: (a) Soil 1; (b) Soil 2; and (c) Soil 3 (Zaliha et al., 2014).

to the formation of the Si-O-Si bond in the geopolymer, increasing its compressive strength (Atmaja et al. 2011). One tetrahedral silica sheet and one octahedral alumina sheet, which are joined together by sharing a common layer of oxygen and hydroxyls,comprise the structure of kaolinite (Heah et al., 2012).

Microstructure Analysis. The micrographs were taken to study the morphological properties of the clays. Figure 10.6 shows the SEM micrograph of the analyzed samples of Soils 1, 2, and 3, which are represented by Figure 10.6a–c. According to Heah et al. (2012), kaolinite morphological features have a plate-like structure that allows the geopolymerization process to occur in such a small surface area, resulting in low reactivity and contributing to the achievement of low compressive strength. A variety of morphological surfaces displayed by quartz due to weathering in terms of chemical dissolution and physical forces, with common features of conchoidal fracture, arc shaped, parallel steps, facet, high and low relief grains, authigenic quartz, rounded and angular form, and solution pits (Wilson, 2020).

Liquid Limit. Alkaline activator was added to the soils with varying solid-to-alkaline activator (S/L) ratios for the liquid limit test. It was demonstrated that the geopolymerization procedure influenced the liquid limit values. Figure 10.7 illustrates the liquid limit values of Soil 2. The sample with the highest liquid limit (51.32%) had a solid-to-alkaline activator (S/L) ratio of 2.0, followed by the control sample with

FIGURE 10.6   SEM micrographs of analyzed samples: (a) Soil 1; (b) Soil 2; and (c) Soil 3.

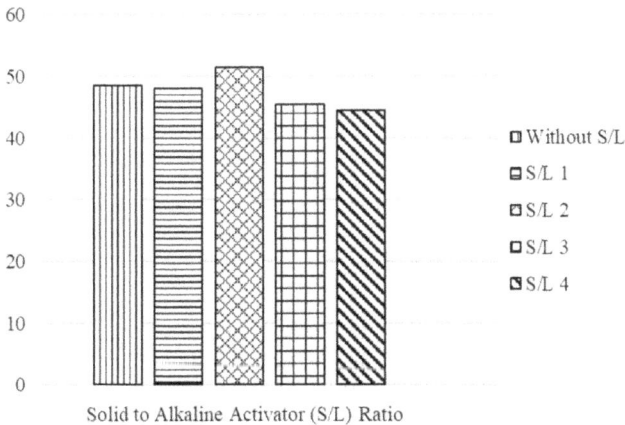

FIGURE 10.7   Liquid limits of Soil 2.

48.69%. Other samples with solid-to-alkaline activator (S/L) ratios of 1.0, 3.0, and 4.0 had 47.98%, 45.62%, and 44.54%, respectively.

Figure 10.8 presented the liquid limit values of Soil 3, which followed the same trend as the graph of Soil 2, with the sample with the highest liquid limit, 61.79%, having a solid-to-alkaline activator (S/L) ratio of 2.0. The liquid limit of the control sample was 54.47%, followed by 54.42% for the sample with a solid-to-alkaline

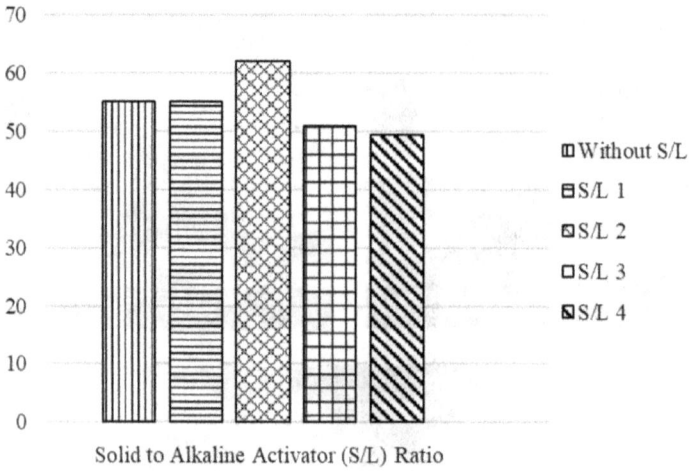

FIGURE 10.8   Liquid limits of Soil 3.

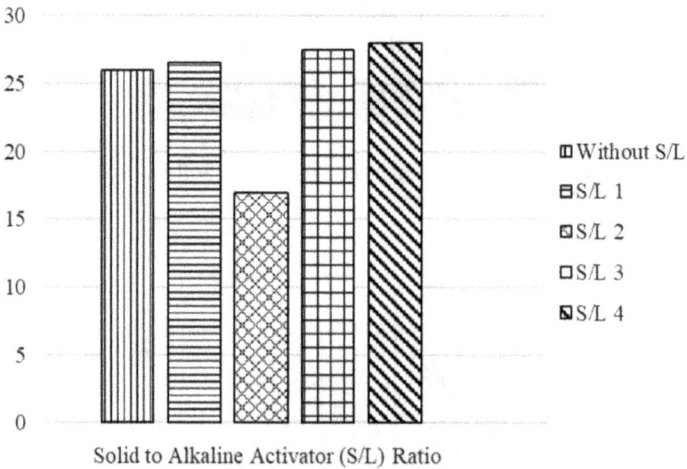

FIGURE 10.9   Plastic limits of Soil 2.

activator (S/L) ratio of 1.0. The sample with a solid-to-alkaline activator (S/L) ratio of 3.0 showed 51.16% liquid limit, and the sample with an S/L ratio of 4.0 showed 49.53% (Syed Zuber et al., 2015).

Plastic Limit. As with the liquid limit test, the geopolymerization process revealed different plastic limit values when alkaline activator was added to soils with a range of solid-to-alkaline activator (S/L) ratios, as presented in Figure 10.9. The sample with the lowest plastic limit, 17.21%, had a solid-to-alkaline activator (S/L) ratio of 2.0. The plastic limit values of the control sample and the sample with an S/L ratio of 1.0 had a small incremental of 26.25% and 26.92%, respectively. The plastic limit values for samples with solid-to-alkaline activator (S/L) ratios of 3.0 and 4.0 were also 27.28% and 28.19%, respectively.

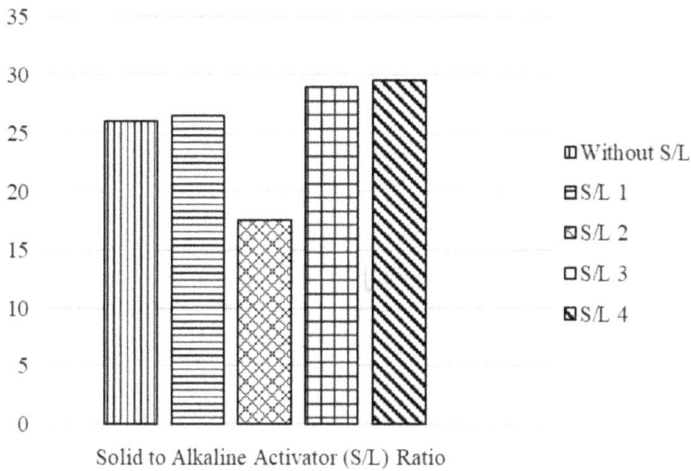

FIGURE 10.10 Plastic limits of Soil 3.

Figure 10.10 showed the plastic limit values of Soil 3, and the sample with the lowest plastic limit (17.70%) had a solid-to-alkaline activator (S/L) ratio of 2.0. The graph clearly followed the same pattern as Soil 2. The control sample indicated a 26.27% plastic limit value, followed by the sample with an S/L ratio of 1.0, 3.0, and 4.0 with incremental values of 26.85%, 28.49%, and 29.13%, respectively.

10.5 CONCLUSIONS

This chapter examined the many geopolymer variants that have been applied to soil stabilization in geotechnical and pavement engineering and focused on the stabilizing processes, engineering qualities, and physico-chemical characteristics of these geopolymers as well as the geopolymerization-treated various soils physical and chemical characteristics. As a conclusion, the strength of soil can be increased by using these materials and techniques in soil stabilization. Overall, the choice of materials used to stabilize the soil will depend on the specific project requirements, soil type, and so on.

An inorganic polymer known as a geopolymer is created when an activator (alkali) and a precursor engage in geopolymerization (Al and Si sources). The activator and precursor employed, as well as the doses and concentrations and the aluminosilicate structures, all have a significant impact on the efficacy of geopolymers. Geopolymerization is the method of geopolymer stabilization in which alumina and silica first dissolve under high pH conditions, then form gel-like complexes, and finally harden. It is important to carefully evaluate and select the appropriate material for soil stabilization to ensure the success of the projects.

The review's conclusions suggest that alkaline binder formulations should be improved in order to lessen their negative effects on the economy and environment. The goal of optimization should be to utilize less NaOH and/or Na_2SiO_3, or even to replace them with inexpensive materials or other wastes with little market value, particularly in solid form, such as rice husk ash asphalt, copper slag, steel slag, ground granulated blast-furnace slag (GGBS), fly ash, and so on. NaOH and/or Na_2SiO_3 manufacturing and transportation both produce considerable amounts of CO_2, which should be reduced as much as possible to slow global warming.

However, there are various difficulties in applying geopolymers for soil stabilization. It is challenging to utilize polymers widely and confidently due to the absence of systematic and independent published research, disorganized labeling of polymers on the market, inadequate criteria for evaluating performances, and inconsistent application rate(s) advised by suppliers. As a result, this study indicates the need for more research in the areas of developing acceptable methodologies for polymer assessment, figuring out how soil polymer mixtures work, analyzing in situ qualities, and addressing durability problems.

REFERENCES

Angriawan, M. (2019). Manufacturing process with various determinants and properties of ordinary portland cement. *Forest Chemicals Review*, 7–12.

Atmaja, L., Fansuri, H., & Maharani, A. (2011). Crystalline phase reactivity in the synthesis of fly ash-based geopolymer. *Indonesian Journal of Chemistry*, *11*(1), 90–95.

Behnood, A. (2018). Soil and clay stabilization with calcium-and non-calcium-based additives: A state-of-the-art review of challenges, approaches and techniques. *Transportation Geotechnics*, 17, 14–32.

Brasse, K., Tracz, T., Zdeb, T., & Rychlewski, P. (2018). Influence of soil-cement composition on its selected properties. *MATEC Web of Conferences*, *163*, 06006.

British Standard-BS 1377-2:1990 (1990). *Methods of Test for Soils for Civil Engineering Purposes: Part 2: Classification Tests*. British Standard.

Çevik, A., Alzeebaree, R., Humur, G., Niş, A., & Gülşan, M. E. (2018). Effect of nano-silica on the chemical durability and mechanical performance of fly ash based geopolymer concrete. *Ceramics International*, 44(11), 12253–12264.

Chindaprasirt, P., Chareerat, T., & Sirivivatnanon, V. (2007). Workability and strength of coarse high calcium fly ash geopolymer. *Cement and Concrete Composites*, 29(3), 224–229. doi:10.1016/j.cemconcomp.2006.11.002

Davidovits, J. (1991). Geopolymers. *Journal of Thermal Analysis*, 37(8), 1633–1656. doi:10.1007/BF01912193

Dhar, S., & Hussain, M. (2019). The strength behaviour of lime-stabilized plastic fibre-reinforced clayey soil. *Road Materials and Pavement Design*, *20*(8), 1757–1778.

Elimbi, A., Tchakoute, H. K., & Njopwouo, D. (2011). Effects of calcination temperature of kaolinite clays on the properties of geopolymer cements. *Construction and Building Materials*, 25(6), 2805–2812. doi:10.1016/j.conbuildmat.2010.12.055

Ghadir, P., & Ranjbar, N. (2018). Clayey soil stabilization using geopolymer and Portland cement. *Construction and Building Materials*, 188, 361–371.

Hamzah, H. N., Al Bakri Abdullah, M. M., Cheng Yong, H., Arif Zainol, M. R. R., & Hussin, K. (2015). Review of soil stabilization techniques: Geopolymerization method one of the new technique. *Key Engineering Materials*, 660, 298–304. doi:10.4028/www.scientific.net/KEM.660.298

Heah, C. Y., Kamarudin, H., Mustafa Al Bakri, A. M., Bnhussain, M., Luqman, M., Khairul Nizar, I., & Liew, Y. M. (2012). Study on solids-to-liquid and alkaline activator ratios on kaolin-based geopolymers. *Construction and Building Materials*, 35, 912–922. doi:10.1016/j.conbuildmat.2012.04.102

Huang, J., Kogbara, R. B., Hariharan, N., Masad, E. A., & Little, D. N. (2021). A state-of-the-art review of polymers used in soil stabilization. *Construction and Building Materials*, 305, 124685.

Ismail, I., Bernal, S. A., Provis, J. L., San Nicolas, R., Hamdan, S., & Van Deventer, J. S. J. (2014). Modification of phase evolution in alkali-activated blast furnace slag by the incorporation of fly ash. *Cement and Concrete Composites*, 45, 125–135. doi:10.1016/j.cemconcomp.2013.09.006

Meek, A. H., Elchalakani, M., Beckett, C. T., & Dong, M. (2021). Alternative stabilized rammed earth materials incorporating recycled waste and industrial by-products: A study of mechanical properties, flexure and bond strength. *Construction and Building Materials*, 277, 122303.

Nnochiri, E. S., Ogundipe, O. M., & Emeka, H. O. (2018). Effects of snail shell ash on lime stabilized lateritic soil. *Malaysian Journal of Civil Engineering*, 30(2). doi:10.11113/mjce.v30n2.477

Prambauer, M., Wendeler, C., Weitzenböck, J., & Burgstaller, C. (2019). Biodegradable geotextiles-An overview of existing and potential materials. *Geotextiles and Geomembranes*, 47(1), 48–59.

Sorsa, A., Senadheera, S., & Birru, Y. (2020). Engineering characterization of subgrade soils of Jimma town, Ethiopia, for roadway design. *Geosciences*, 10(3), 94.

Syed Zuber, S. Z., Al Bakri Abdullah, M. M., Hussin, K., Ahmad, F., & Binhussain, M. (2015). The influence of geopolymerization process on liquid and plastic limits of soils. *Applied Mechanics and Materials*, 754–755, 886–891. doi:10.4028/www.scientific.net/AMM.754-755.886

Wang, T., Wang, J., Hou, X., & Xiao, F. (2021). Effects of SARA fractions on low temperature properties of asphalt binders. *Road Materials and Pavement Design*, 22(3), 539–556.

Wilson, M. J. (2020). Dissolution and formation of quartz in soil environments: A review. *Soil Science Annual*, 71, 3–14.

Yuan, J., Li, L., He, P., Chen, Z., Lao, C., Jia, D., & Zhou, Y. (2021). Effects of kinds of alkali-activated ions on geopolymerization process of geopolymer cement pastes. *Construction and Building Materials*, 293, 123536.

Zaliha, S. Z. S., Al Bakri, A. M. M., Kamarudin, H., & Fauziah, A. (2014). Characterization of soils as potential raw materials for soil stabilization application using geopolymerization method. *Materials Science Forum*, 803, 135–143. doi:10.4028/www.scientific.net/MSF.803.135

11 Geopolymer Bricks
A Case Study

Wan Mastura Wan Ibrahim,
Laila Mardiah Deraman,
Mohd Mustafa Al Bakri Abdullah,
and Romisuhani Ahmad
Universiti Malaysia Perlis

Puput Risdanareni
Universitas Brawijaya

11.1 INTRODUCTION

The huge demand from the construction industry due to the expanded population has entailed the need for sustainable building constituents, especially bricks (Anastasiades et al., 2021). Bricks are extensively used in building and construction materials worldwide. Bricks are usually used in the structure of buildings as a construction wall, facing paving, perimeter, garden wall and flooring (Sánchez-Garrido et al., 2022). Generally, bricks are produced in various types, such as cement brick, concrete brick and clay brick. These bricks are constructed using different materials, sizes and curing times (Al-Fakih et al., 2019). The global annual production of bricks is currently about 1,391 billion units, and the demand for bricks is probably to continue increasing (Sutcu et al., 2015). Conventional bricks were manufactured from clay with a high temperature of kiln firing (900°C–1,000°C) or from ordinary Portland cement (OPC) concrete (Wan Ibrahim et al., 2015).

Clay bricks integrate around 2.0 kWh of energy and release about 0.41 kg of carbon dioxide (CO_2) per brick (Ahmari and Zhang, 2012; Murekar, 2017). The use of other raw materials apart from clay, such as waste materials, is something that needs to be emphasized to protect the clay resource and the environment. Concrete bricks are produced from OPC and aggregates. Production of OPC has led to the release of a significant amount of carbon dioxide (CO_2) and greenhouse gases (GHGs), generating global warming. With every ton of cement produced, almost a tonne of CO_2 is emitted (Naqi & Jang, 2019).

Geopolymer, which seems similar to natural zeolitic materials with an amorphous phase, is a capable new formula of inorganic polymer material that could be another good substitution for OPC in the manufacturing of cement brick. The low-temperature geopolymeric setting (LTGS) was the first patented geopolymer brick produced in 1982 in France. Geopolymer bricks need about eight times less energy, are more affordable and require less equipment in production as compared to clay

DOI: 10.1201/9781003390190-11

bricks, which are fired at 1,000°C in a kiln (Murmu & Patel, 2018). Geopolymeric bricks are considered as a new technology for eco-sustainable masonry units because they possess good mechanical and physical properties and increase the possibilities of recycling surplus material into valuable products, especially construction material (Buchwald et al., 2016). As such, for a more sustainable environment, many researchers completed their study with other materials to produce bricks using by-product materials.

11.2 GEOPOLYMER BRICKS

In the construction of bricks and other building materials, an important raw material used is mainly composed of cement, aggregates and water. The most popular binder is Portland cement, which releases a considerable amount of carbon dioxide during its manufacturing (Aprianti, 2017). Therefore, to reduce the GHG emission from the manufacturing, there is a need to reduce or avoid traditional cement as possible. Geopolymer technology is one in which the complete elimination of cement is achieved without compromising strength and durability (Kamseu et al., 2010; Valente et al., 2022).

Geopolymer is the chemical reaction between aluminosilicate reactive material and strong alkaline solutions such as sodium hydroxide solution and sodium silicate solution that yields amorphous aluminosilicate material (Amran et al., 2020). Geopolymers have been increasing in interest, study and utilization worldwide for some decades. Further inspiration for discovering this alternative is attributed to better fire and acid resistance, high compressive strength and high flexural strength (Li et al., 2013). Geopolymers are usually produced using either metakaolin or fly ash as a precursor and have flow structures initiating from the condensation process of tetrahedral aluminosilicate units of different Si/Al ratios, such as $(Al–O–Si–O–Si–O–)$ M^+, $(Al–O–Si–O–)M^+$, $(Si–O–Al–O–Si–O–Si–O–)M^+$, and so on, where M^+ is an alkali ion, which generally comes from Na^+ or K^+ ion, which balances the charge of the tetrahedral Al (Tanyildizi, 2021).

The basic principle and simplified structure of the formation of fly ash-based geopolymer is the breakdown of aluminosilicate in the fly ash and then the poly-condensation process. The process begins with the reactions between fly ash and alkaline solution and condensation between the resultant Si^{4+} and Al^{3+} classes, followed by other complex nucleation of monomers, followed by oligomerization and finally the polymerization process, which leads to a new aluminosilicate-based polymer with a different amorphous three-dimensional network structure (Zhuang et al., 2016).

A previous study conducted on bricks made of fly ash, bottom ash and alkali solution confirmed the feasibility of brick production (Madani et al., 2020). In a study, bricks of strength ranging from 5 to 60 MPa were successfully prepared by using Class F fly ash as a raw material, sodium silicate and sodium hydroxide solutions with a forming pressure of 30 MPa and cured at elevated temperatures (Arioz et al., 2013). Besides, several studies have introduced low-cost clays (white, grey and red clay) for the production of geopolymer bricks (Mohsen and Mostafa, 2010). These bricks used sodium hydroxide (NaOH) and sodium silicate (Na_2SiO_3) as alkaline activators and were activated by calcinations at 700°C for 2 hours. The mixture was moulded into cylindrical specimens under a moulding pressure of 15 MPa in a steel mould and cured for 24 hours. From the research, the curing process and alkaline activator were significant aspects to getting better mechanical properties for geopolymer bricks (Mohsen and Mostafa, 2010).

Mastura et al. (2014) revealed that the utilization of fly ash as a source material for geopolymer brick production appears to be a reasonable solution that allows for the conservation of natural resources, reduces further pollution and saves the environment. Fly ash-based geopolymer bricks were manufactured by varying the ratio of fly ash to sand (1:2–1:5, by mass) and using curing time (1–24 hours) and curing temperature (80°C). The results noted for compressive strength up to 20.3 MPa were obtained by curing at 70°C for a period of 24 hours at 60 days of ageing (Mastura et al., 2014). For other researchers in kaolin-based geopolymer bricks, the binder of alkaline activators also includes sodium hydroxide (NaOH) and sodium silicate (Na_2SiO_3) because they are easy to handle and inexpensive. The researcher used three mix designs: ratio of sand/kaolin (6–10), by mass; ratio of kaolin/alkali solution (0.5–1.5), by mass; and ratio of sodium silicate/sodium hydroxide (0.2–0.4), by mass (Muhammad Faheem et al., 2013). The results obtained met the requirements of ASTM C129 (2017) and ASTM C90 (2017).

11.3 IBS BRICKS

The Industrialized Building System (IBS) is a key to futuristic construction. IBS can be defined as a construction technique in which the components are manufactured in a controlled environment (or off-site), transported, positioned and assembled into a structure with minimal additional site work (Norazlin et al., 2021). On-site precasting consists of floor and roof slabs in situ, whereas off-site fabrications of some or all components of buildings are cast off-site at fabrication yards or factories. With the transfer of construction operations to factories or fabrication yards, good quality components have been mass produced and delivered to the construction sites in economically large loads (Suresh et al., 2012).

The adoption of IBS offers valuable advantages such as the reduction of unskilled workers, less waste, a smaller volume of site materials, improved site cleanliness, reduced environmental pollution, better quality control, promotion of site safety practices and reduced construction time. Structural classification under the IBS roadmap falls into five main categories: precast concrete (PC) framing/panel and box systems, formwork systems, steel framing systems, prefabricated timber framing systems and block-work systems. Among the five structural categories, three systems are concrete-related because concrete is by far the most preferred material in Malaysia. PC framing systems, panel systems and box systems that utilize prefabrication technology fit the definition of IBS as compared to formwork systems, which still involve on-site activities (Nurul Asra et al., 2015).

The use of IBS will overcome the issues of repetitive parts of the building but difficult, time-consuming and costly labour at the site. At the same time, IBS also involves on-site casting using innovative and clean mould technologies such as steel, aluminium and plastic. The philosophy of shifting from the traditional practice of laying brick and mortar into mechanized and automated prefabricated works makes IBS an innovative construction technique and simultaneously forms pressure for the Malaysian Construction Industry to accomplish the implementation target (Izatul Laili et al., 2015).

11.4 MIX DESIGN, PROCESSING AND CURING PROCESS

11.4.1 BOTTOM ASH WITH REPLACEMENT FLY ASH AS GEOPOLYMER BRICKS

The materials, which are bottom ash, fly ash and alkali solution ($Na_2SiO_3 + NaOH$), were weighed. Alkali solution is added gradually with solid materials (fly ash and bottom ash). The mixing process is done through a dry pan mixer for producing geopolymer brick samples. The solid material or geopolymer raw materials need to be mixed first for 5 minutes before adding an alkaline activator. The mixed materials were then weighed at approximately 2.5 kg, poured into the mould and compressed with a pressure of 10 MPa. Curing temperature and curing time were fixed as selected parameters for all geopolymer brick samples.

The bricks were cured in the oven at 80°C for 24 hours after being removed from the mould. The geopolymer brick samples were kept at room temperature for 7 days prior to being tested. To determine the optimum value of mix design for non-loading application geopolymer bricks, the specimen was compressed, density tested and water absorption tested after 7 days. In the study on the percentage of fly ash addition to geopolymer bricks, Na_2SiO_3-to-NaOH ratio and solid-to-liquid ratio were fixed at 2.5 and 2.0, respectively. The molarity (M) of NaOH was fixed at 12 M for each sample. These ratios were used to obtain good workability and strength for geopolymer bricks. The details of the mix designs are shown in Table 11.1. The mass of NaOH and Na_2SiO_3 were fixed for all samples.

11.4.2 KAOLIN GEOPOLYMER BRICKS

The mix design of kaolin geopolymer bricks is studied based on the ratio of sand to kaolin. The analysis is represented in Table 11.2. Sand-to-kaolin ratios were set at

TABLE 11.1
Mix Design for Effect on Percentage of Fly Ash Addition

Description	Mix Design
Ratio of Na_2SiO_3 to NaOH	2.50
Ratio of solid to liquid	2.00
Mass of NaOH (g)	357.14 g
Mass of Na_2SiO_3 (g)	892.86 g

8, 7, 6, 5 and 4. The NaOH concentration used is 8 M, kaolin to activator is 1.0 and Na$_2$SiO$_3$ to NaOH is 0.3. The kaolin-based geopolymer brick samples were cured in the oven at 80°C for 24 hours.

11.4.3   POBA Geopolymer Bricks/IBS Bricks

The mix design of POBA geopolymer bricks/IBS bricks was discussed in this section. To produce 14 M of NaOH, NaOH pellets were diluted in distilled water. The NaOH solution mixed with sodium silicate fixed the ratios of POBA to alkaline activator and sodium silicate to NaOH at 1.5 and 2.5, respectively. Then, the ratio of POBA to sand used was 1:3. Details of the mix design were revealed in Table 11.3. In addition, sand and POBA were mixed in the pan mixer for 3 minutes. The mixing process continued for 5 minutes after an alkaline activator was added to the mixer. The mixture was moulded and cured in the oven at 80°C for 24 hours of ageing.

TABLE 11.2
Details Mix Proportions with Different Sand-to-Kaolin Ratio (Muhammad Faheem et al., 2016)

Types			Kaolin		
Molarity of NaOH			8 M		
Weight for 1 brick (g)			2,500		
No. of bricks			1		
Sand-to-kaolin ratio	8.00	7.00	6.00	5.00	4.00
Mass of sand	2222.22	2187.50	2142.86	2083.33	2000.00
Mass kaolin (g)	277.78	312.50	357.14	416.67	500.00
Kaolin-to-activator ratio			1.00		
Mass of activator (g)	277.78	312.50	357.14	416.67	500.00
Na$_2$SiO$_3$-to-NaOH ratio			0.30		
Mass NaOH solution (g)	213.68	240.39	274.73	320.51	384.61
Mass Na$_2$SiO$_3$ solution (g)	64.10	72.12	82.42	96.15	115.39

TABLE 11.3
Details Mix Design of Geopolymer/IBS Bricks
(Zarina et al., 2016)

	Geopolymer Bricks	Geopolymer IBS Bricks
POBA:sand ratio	1:3	
POBA (g)	600.00	750.00
Sand (g)	1800.00	2250.00
NaOH (g)	114.20	142.90
Na$_2$SiO$_3$ (g)	285.80	357.10

11.5 MECHANICAL AND PHYSICAL PROPERTIES OF GEOPOLYMER BRICKS

11.5.1 BOTTOM ASH WITH REPLACEMENT FLY ASH AS GEOPOLYMER BRICKS

In the study of the effect of the use of bottom ash on the percentage of fly ash replacement, the percentage of fly ash should not exceed 50%. The compressive strength, density and water absorption of these geopolymer bricks are shown in Figures 11.1–11.3. The best strength is observed at 40% fly ash added with 6.98 MPa, while the lowest strength (1.38 MPa) was recorded at no fly ash added. The compressive strength increased gradually from 1.38 MPa (0%) to 2.56 MPa (10%), 4.92 MPa (20%), 6.42 MPa (30%) and 6.98 MPa (40%), respectively.

Generally, increasing the percentage of fly ash would increase the compressive strength of geopolymer bricks. This is due to the irregular shape and size of the bottom ash particles, and the coarse surface contributed to the high inter-particle friction, which reduced the workability of the sample when too much bottom ash was added. Besides, the properties of bottom ash could act as a water tank that keeps the water and releases it back into the mixture during the mixing process (Morla et al., 2021).

The densities of geopolymer bricks with different percentages of fly ash replacement are shown in Figure 11.2. The graph shows an ascending order in the range 1,289–1,399 kg/m³. The highest density recorded was 40% (1,399 kg/m³) of fly ash, while the lowest density showed 0% fly ash at 1,289 kg/m³. The density of geopolymer bricks generally depends on their compressive strength. The increment in density happened with the increment in fly ash content. The ability of fly ash to fill up the voids and increase the tendency of air to become entrapped on the surface and structure of geopolymer bricks is increased by the coarse size of the bottom ash. In addition, the pores in the bottom ash were filled with the percentage of fly ash added, thereby increasing the density and strength of geopolymer bricks.

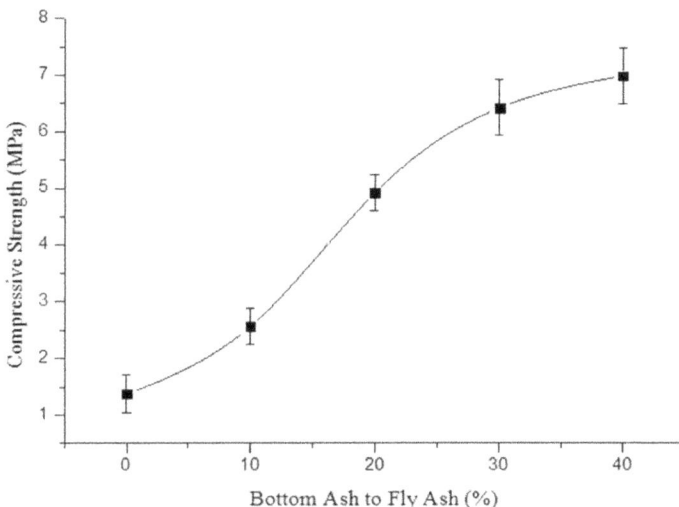

FIGURE 11.1 Compressive strength of replacement fly ash.

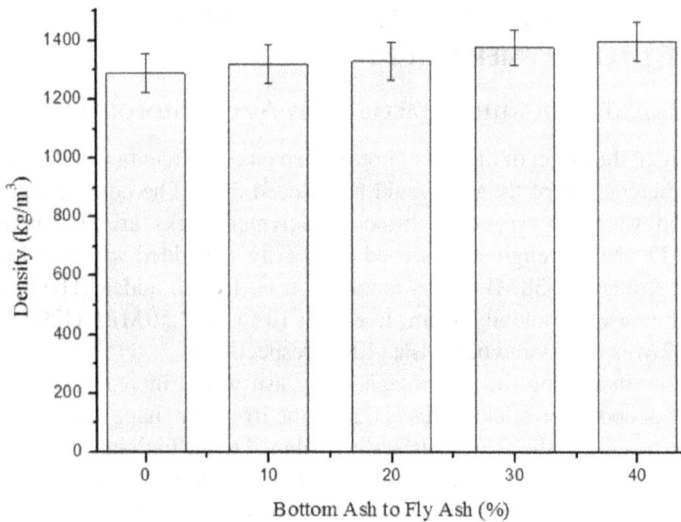

FIGURE 11.2 Density of percentage of fly ash addition.

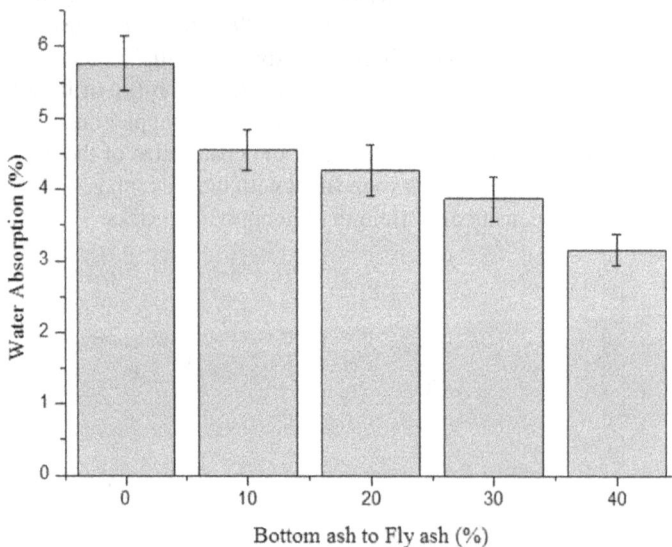

FIGURE 11.3 Water absorption of percentage fly ash addition.

The results of water absorption for the percentage of fly ash replacement geopolymer bricks are imprecise in Figure 11.3. The higher absorption of water is revealed at 0 additions of fly ash, which is 5.76%, and the lowest at 40% additions of fly ash (3.16%). The effect of the water absorption can be seen in the reduced compressive strength of this geopolymer brick. The lowest water absorption percentages are observed at 40% of fly ash, as the best amount of fly ash is presented during the geopolymerization process. The samples produced undergo complete geopoly-merization during the curing period, forming a strong geopolymer paste. Giogetti Deutou Nemaleu et al. (2021) mentioned that the water absorption mainly depends on the capillary pore and the artificial pores in the geopolymer brick sample to deter-mine the compressive strength and density. Besides that, the lower water absorption shows higher resistance to water penetration and less environmental damage.

11.5.2 Kaolin Geopolymer Bricks

The compressive strength results of kaolin-based geopolymer bricks with various ratios of sand/kaolin are shown in Figure 11.4. The increase in the ratio of sand/kaolin increased the compressive strength of geopolymer bricks. The lowest strength was observed at a ratio of sand/kaolin of 4, by mass, and the highest strength was achieved at a ratio of sand/kaolin of 8, by mass. The strength of kaolin-based geo-polymer bricks increases because of the presence of calcium in sand, which affects the hardening process of geopolymerization and enhances the early strength of geo-polymer materials (Ishwarya et al., 2019).

Figure 11.5 shows the density values of kaolin-based geopolymer bricks with various ratios of sand/kaolin. The density values are in the range between 1,636 and 2,242 kg/m³. The highest density value was presented at the samples with a ratio of sand/kaolin of 8, by mass, and the lowest (1,636 kg/m³) density appeared at the samples with a ratio of sand/kaolin of 4, by mass. This is because the amount of sand used increased and improved the density of geopolymer bricks. The lowest density of geopolymer bricks could be considered as lightweight bricks, according to ASTM C129 (2017).

The water absorption of kaolin-based geopolymer bricks with different ratios of sand/kaolin, by mass is presented in Figure 11.6. The lowest water absorption (9.20%) was illustrated at a sand/kaolin ratio of 8. The highest water absorption (21.65%) was found in a sample with a sand/kaolin ratio of 4. This result demonstrates that the

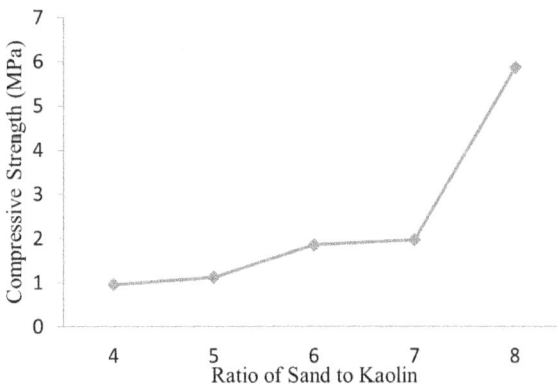

FIGURE 11.4 Compressive strength of kaolin-based geopolymer bricks with different sand-to-kaolin ratios (Muhammad Faheem et al., 2016).

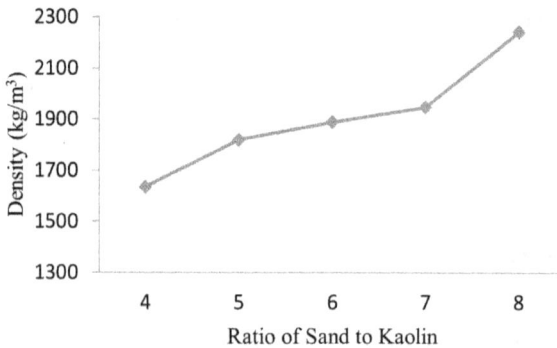

FIGURE 11.5 Density of kaolin-based geopolymer bricks with different sand-to-kaolin ratios (Muhammad Faheem et al., 2016).

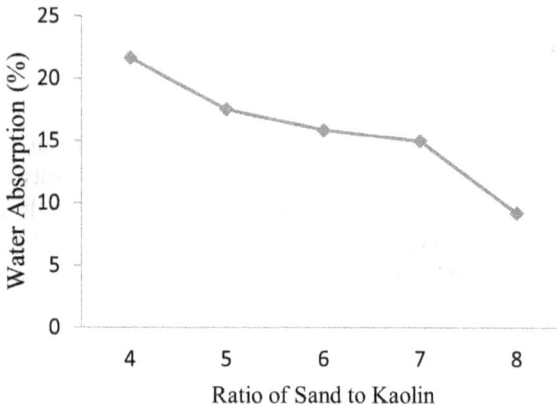

FIGURE 11.6 Percentage of water absorption of kaolin-based geopolymer bricks with different sand-to-kaolin ratios (Muhammad Faheem et al., 2016).

increasing amount of sand caused an increase in the density of geopolymer bricks, which contributed to the decrease in water absorption value.

11.5.3 POBA GEOPOLYMER BRICKS/IBS BRICKS

The geopolymer bricks could also be produced by using palm oil boiler ash (POBA) as a source material. The POBA ash is known as a waste material, with a high amount of silica (SiO_2) and CaO. The compressive strength of POBA-based geopolymer bricks was tested at the ages of the first, third, seventh, 28th and 60th days, as shown in Figure 11.7. The strength of POBA-based geopolymer bricks shows an increment with ageing. The highest strength (16.10 MPa) was achieved at the 60th day of ageing, and the lowest strength (4.20 MPa) was found at the first day of ageing. The strength results obtained have met the requirements for non-loadbearing bricks according to ASTM C129. As the ageing days increases, they enhance the formation of an aluminosilicate network and improve the strength of POBA-based geopolymer bricks.

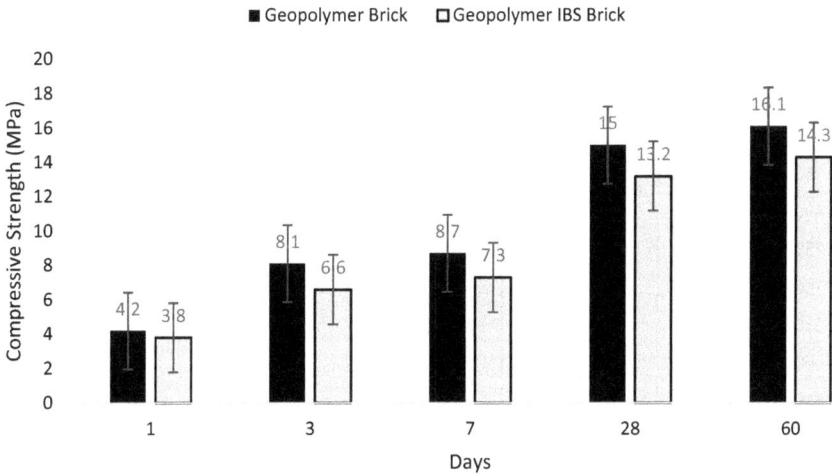

FIGURE 11.7 Compressive strength of POBA-based geopolymer/IBS bricks with different ageing days (Zarina et al., 2016).

The compressive strength results for the IBS bricks are also depicted in Figure 11.7. The highest strength of 14.30 MPa was attained at the 60th day of ageing. The compressive strength of IBS bricks met the requirement for non-loadbearing bricks as stated in ASTM C129 (2017). The strength of IBS bricks increases withincreasing ageing days. Nevertheless, the strength of IBS bricks and POBA-based geopolymer bricks has a slight difference due to the presence of grooves and tongues on the surface of IBS bricks (Zarina et al., 2016).

The density of POBA-based geopolymer bricks with different ageing days is shown in Figure 11.8. The average densities obtained are in the range of 1,615 and 1,742 kg/m³. The geopolymer bricks could be categorized as medium weight non-loadbearing bricks according to ASTM C129 (2017) based on their highest density of 1,742 kg/m³. The highest density acquired is due to the complete geopolymerization process, which contributed to the denser geopolymer and higher strength. This is because, with increasing ageing days, the rate of the geopolymerization process increases, resulting in denser and less porous geopolymer structures (Patel & Shah, 2018). Figure 11.8 also shows the density of IBS bricks in the range of 1,792–1,894 kg/m³. The IBS bricks could be classified as medium weight bricks according to the ASTM C129 (2017) standard. The density and compressive strength of geopolymer IBS bricks have a positive relationship, where higher density bricks have better strength properties.

Figure 11.9 shows the water absorption values for POBA-based geopolymer bricks with different ageing days. A sample cured at 60 days gives the lowest (6.80%) value of water absorption, which contributed to the highest compressive strength. The highest water absorption (12.20%) is detected on the first day of ageing. Curing time plays an important role in producing the compact structure of geopolymer bricks because the aluminosilicate monomers are continuously reacted to each other until a complete reaction is achieved. The samples cured at the 60th day of ageing produced a denser structure and a value lower than the requirement stated in the standard specification ASTM C90 (2017).

■ Geopolymer Brick   □ Geopolymer IBS Brick

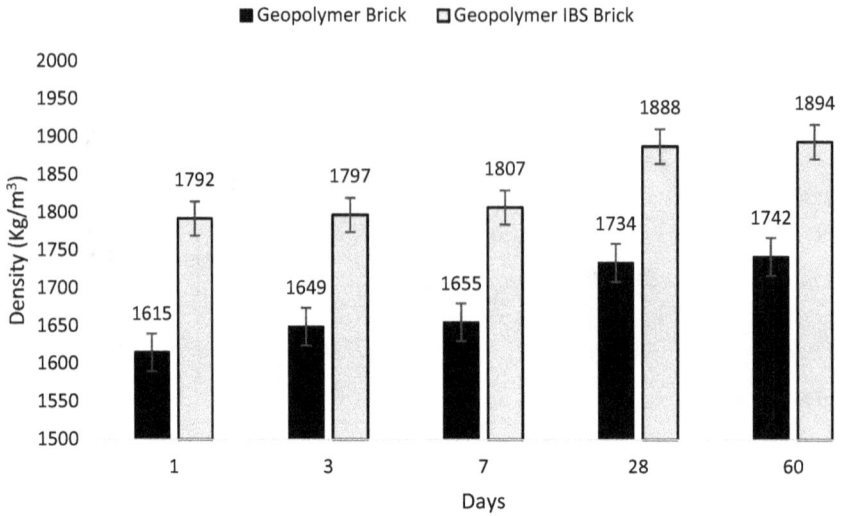

FIGURE 11.8 Density of POBA-based geopolymer/IBS bricks with different ageing days (Zarina et al., 2016).

■ Geopolymer Brick   □ Geopolymer IBS Brick

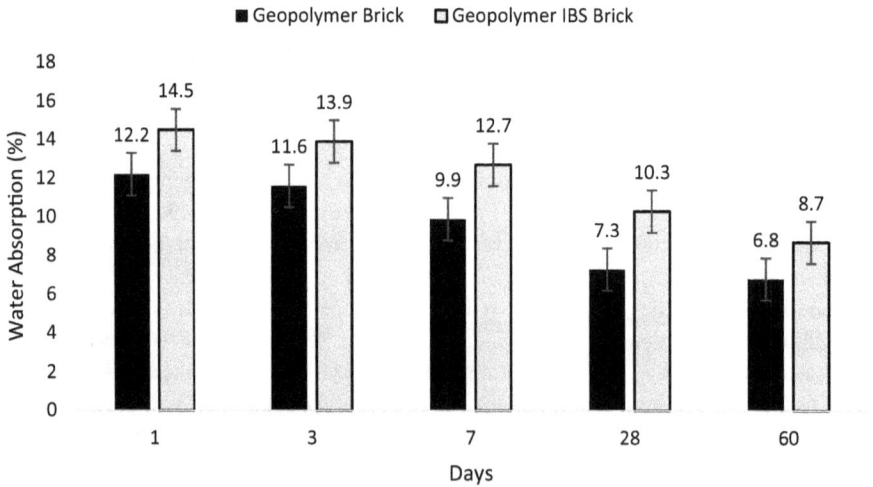

FIGURE 11.9 Water absorption of geopolymer/IBS bricks with different ageing days (Zarina et al., 2016).

Figure 11.9 shows the water absorption of IBS geopolymer bricks decreases with increasing ageing days from the first to 60th days. The lowest percentage of the sample at the 60th day is due to the better resistance to water permeation and reduced environmental destruction (Mohsen and Mostafa, 2010). The water absorption of IBS bricks is in the range of 8.70%–14.50%. The lowest value of water absorption is achieved due to the complete geopolymerization process, which produces denser bricks and reduces water absorption.

11.6 CONCLUSION

In conclusion, geopolymer bricks are a promising alternative to traditional bricks made from clay or cement. Geopolymer bricks are eco-friendly, as they are made from industrial waste materials or natural resources and have a lower carbon footprint compared to conventional bricks. They also exhibit good mechanical properties, such as high strength and durability, making them suitable for various construction applications. The density, water absorption and compressive strength of geopolymer brick discussed in this chapter meet the requirements for a non-loadbearing concrete masonry unit according to ASTM C129 (2017).

Geopolymer bricks represent a promising advancement in sustainable construction practices, and their continued development and adoption have the potential to significantly reduce the environmental impact of construction. However, further research, development and market acceptance are needed to overcome existing challenges and fully realize their potential. With continued advancements in technology and increased awareness of environmental concerns, geopolymer bricks could become a mainstream construction material in the future, contributing to more sustainable and eco-friendly building practices.

Overall, geopolymer bricks are a promising technology with potential benefits for the environment, the construction industry and society as a whole. Further research and development in this field will likely contribute to their wider adoption and integration into mainstream construction practices in the future. As with any emerging technology, continued research, innovation and market acceptance will be key to unlocking the full potential of geopolymer bricks as a sustainable building material.

11.7 FUTURE STUDY

Further studies are required in order to improve, spread and explore the present work for the development of geopolymer bricks, such as:

a. It is recommended to expand the mix design and mix proportion by using other types of by-product material in place of sand to observe the physical properties and strength.
b. Further study on the mechanical properties benefits of the bricks application, such as the initial rate of suction, fire resistance, efflorescence and acid resistance.
c. Investigate the geopolymer bricks under different conditions, both at elevated temperatures and at room temperature.

REFERENCES

Ahmari, S., & Zhang, L. (2012). Production of eco-friendly bricks from copper mine tailings through geopolymerization. *Construction and Building Materials*, 29, 323–331.
Al-Fakih, A., Mohammed, B. S., Liew, M. S., & Nikbakht, E. (2019). Incorporation of waste materials in the manufacture of masonry bricks: An update review. *Journal of Building Engineering*, 21, 37–54.

Amran, Y. H. M., Alyousef, R., Alabduljabbar, H., & El-Zeadani, M. (2020). Clean production and properties of geopolymer concrete: A review. *Journal of Cleaner Production*, 251, 119679.

Anastasiades, K., Goffin, J., Rinke, M., Buyle, M., Audenaert, A., & Blom, J. (2021). Standardisation: An essential enabler for the circular reuse of construction components? A trajectory for a cleaner European construction industry. *Journal of Cleaner Production*, 298, 26864.

Aprianti S. E. (2017). A huge number of artificial waste material can be supplementary cementitious material (SCM) for concrete production: A review Part II. *Journal of Cleaner Production*, 142, 4178–4194.

Arioz, E., Arioz, Ö., & Kockar, Ö. M. (2013). The effect of curing conditions on the properties of geopolymer samples. *International Journal of Chemical Engineering and Applications*, 4(6), 423–426.

ASTM C90. (2017) *Standard Specification for Loadbearing Concrete Masonry Units* (Vol. 4.05). ASTM International, West Conshohocken, PA, USA.

ASTM C129. (2017). *Standard Specification for Nonloadbearing Concrete Masonry Units* (Vol. 4.05). ASTM International, West Conshohocken, PA, USA.

Bakri, A. M. Mustafa Al, H. Kamaruddin, M. Bnhussain, I. K., Nizar, W. I. W. M. (2011). Mechanism and chemical reaction of fly ash geopolymer cement: A review abstract. *Journal of Asian Scientific Research*, 1(5), 247–253.

Buchwald, E. (2016). A hierarchical terminology for more or less natural forests a hierarchical terminology for more or less natural forests in relation to sustainable management and biodiversity conservation. *Food and Agriculture Organization of the United Nations, Proceedings. Third Expert Meeting on Harmonizing Forest -related Definitions* (11–19). Rome.

Giogetti Deutou Nemaleu, J., Rodrigue Kaze, C., Valdès Sontia Metekong, J., Adesina, A., Alomayri, T., Stuer, M., & Kamseu, E. (2021). Synthesis and characterization of eco-friendly mortars made with RHA-NaOH activated fly ash as binder at room temperature. *Cleaner Materials*, 1, 100010.

Ishwarya, G., Singh, B., Deshwal, S., & Bhattacharyya, S. K. (2019). Effect of sodium carbonate/sodium silicate activator on the rheology, geopolymerization and strength of fly ash/slag geopolymer pastes. *Cement and Concrete Composites*, 97, 226–238.

Izatul Laili, J., Faridah, I., & Abdul Rashid, A. Z. (2015). Public participation: Enhancing public perception towards IBS implementation. *Procedia—Social and Behavioral Sciences*, 168, 61–69.

Li, X., Ma, X., Zhang, S., & Zheng, E. (2013). Mechanical properties and microstructure of class C fly ash-based geopolymer paste and mortar. *Materials*, 6(4), 1485–1495.

Madani, H., Ramezanianpour, A. A., Shahbazinia, M., & Ahmadi, E. (2020). Geopolymer bricks made from less active waste materials. *Construction and Building Materials*, 247, 118441.

Mastura, W. A. N., Ibrahim, W. A. N., Mustafa, M., Bakri, A. L., Sandu, A. V., Hussin, K., Sandu, I. G., Ismail, K. N., Kadir, A. A., & Binhussain, M. (2014). Processing and characterization of fly ash-based geopolymer bricks. *Revista de Chimie*, 11, 1–6.

Mohsen, Q., & Mostafa, N. Y. (2010). Investigating the possibility of utilizing low kaolinitic clays in production of geopolymer bricks. *Ceramics—Silikaty*, 54(2), 160–168

Morla, P., Gupta, R., Azarsa, P., & Sharma, A. (2021). Corrosion evaluation of geopolymer concrete made with fly ash and bottom ash. *Sustainability (Switzerland)*, 13(1), 1–16.

Muhammad Faheem, M. T., Al Bakri, A. M. M., Ghazali, C. M. R., Kamarudin, H., Izzat, A. M., & Abdullah, A. (2013). New processing method of kaolin-based geopolymer brick by using geopolymer brick machine. *Key Engineering Materials*, 594–595, 406–410.

Muhammad Faheem, M. T. (2016). Physical and Mechanical Properties of Kaolin Based Geopolymer Brick. Master Thesis, Universiti Malaysia Perlis, Malaysia.

Murekar, P. N. R. (2017). Using waste material for making light weight bricks. *International Conference on Recent Trends in Engineering Sciences and Technology (ICRTEST 2017)*, 5(1), 467–470.

Murmu, A. L., & Patel, A. (2018). Towards sustainable bricks production: An overview. *Construction and Building Materials*, 165, 112–125.

Naqi, A., & Jang, J. G. (2019). Recent progress in green cement technology utilizing low-carbon emission fuels and raw materials: A review. *Sustainability (Switzerland)*, 11(2), 537

Norazlin, M. S., Muhammad Mursyid, H., Noorlinda, A., & Mohd Hafiz, S. (2021). Industrialized building system (IBS): Challenges in implementing interlocking mortarless blocks (IMB) system for housing projects. *International Journal of Social Science And Human Research*, 4(6), 1419–1425.

Nurul Asra, A. R., Zainab, M. Z., Suryani, A., & Noorsaidi, M. (2015). Maximization of IBS elements at wet areas in solving leaking problems and promoting better quality control. *Procedia—Social and Behavioral Sciences*, 202, 424–435.

Patel, Y. J., & Shah, N. (2018). Enhancement of the properties of ground granulated blast furnace slag based self compacting geopolymer concrete by incorporating rice husk ash. *Construction and Building Materials*, 171, 654–662.

Sánchez-Garrido, A. J., Navarro, I. J., & Yepes, V. (2022). Multi-criteria decision-making applied to the sustainability of building structures based on modern methods of construction. *Journal of Cleaner Production*, 330, 129724.

Suresh, K. L., Joy, J. P., Mohd Raihan, T., & Mazlin, M. (2012). Construction waste minimization comparing conventional and precast construction (Mixed System and IBS) methods in high-rise buildings: A Malaysia case study. *Resources, Conservation and Recycling*, 68, 96–103.

Sutcu, M., Alptekin, H., Erdogmus, E., Er, Y., & Gencel, O. (2015). Characteristics of fired clay bricks with waste marble powder addition as building materials. *Construction and Building Materials*, 82, 1–8.

Tanyildizi, H. (2021). Predicting the geopolymerization process of fly ash-based geopolymer using deep long short-term memory and machine learning. *Cement and Concrete Composites*, 123, 104177.

Valente, M., Sambucci, M., Chougan, M., & Ghaffar, S. H. (2022). Reducing the emission of climate-altering substances in cementitious materials: A comparison between alkali-activated materials and Portland cement-based composites incorporating recycled tire rubber. *Journal of Cleaner Production*, 333, 130013.

Wan Ibrahim, W. M., Hussin, K., al Bakri Abdullah, M. M., Abdul Kadir, A., & Binhussain, M. (2015). A review of fly ash-based geopolymer lightweight bricks. *Applied Mechanics and Materials*, 754–755, 452–456.

Zarina, Y., Kamarudin, H., Al Bakri Abdullah, M. M., Ismail, K. N., & Razak, R. A. (2016). Strength, density and water absoprtion of palm oil boiler ash (POBA) geopolymer brick/ IBS brick. *Key Engineering Materials*, 673, 21–28.

Zhuang, X. Y., Chen, L., Komarneni, S., Zhou, C. H., Tong, D. S., Yang, H. M., & Wang, H. (2016). Fly ash-based geopolymer: Clean production, properties and applications. *Journal of Cleaner Production*, 125, 253–267.

Index

For Product Safety Concerns and Information please contact our EU
representative GPSR@taylorandfrancis.com
Taylor & Francis Verlag GmbH, Kaufingerstraße 24, 80331 München, Germany

www.ingramcontent.com/pod-product-compliance
Lightning Source LLC
Chambersburg PA
CBHW070726220326
41598CB00024BA/3325